EMBARKING ON ADVENTUROUS JOURNEYS THROUGH SOME SUBJECTS IN PURE MATHEMATICS

Part I

First Edition

$$\begin{cases} ax^2+bx+c=0 \quad (a\neq 0) \\ x_{k+1}=-\dfrac{b}{2a}\left(1+\cosh\left[\dfrac{1}{2}\cosh^{-1}\left(\dfrac{b^2-8ac}{b^2}\right)+i\pi k\right]\right) \text{ for } k=0,1. \end{cases}$$

$$\begin{cases} ax^3+bx^2+cx+d=0 \quad (a\neq 0) \\ x_{k+1}=-\dfrac{b}{3a}+\dfrac{2}{3a}\sqrt{b^2-3ac}\cdot\cosh\left(\dfrac{1}{3}\cosh^{-1}\left(\dfrac{9abc-2b^3-27a^2d}{2(b^2-3ac)^{3/2}}\right)+\dfrac{i2\pi k}{3}\right), \ k=0,1,2. \end{cases}$$

$$\begin{cases} ax^7+bx^6+cx^4+\dfrac{c^2}{4b}x^2+\dfrac{c^3}{56b^2}=0 \quad (a\neq 0) \\ x_{k+1}=-\dfrac{\sqrt{c}}{2\sqrt{2b}\sinh\left(\dfrac{1}{7}\sinh^{-1}\left(\dfrac{7a\sqrt{c}}{2b\sqrt{2b}}\right)+\dfrac{i2\pi k}{7}\right)}, \\ \qquad\qquad\qquad k=0,1,\dots,6. \end{cases}$$

$$\begin{cases} x^8+cx^5+dx^4+ex^3+\dfrac{e^4}{c^4}=0 \\ x_{1,\dots,8}=\dfrac{\beta_{k+1}e}{2c}\pm\dfrac{1}{2c}\sqrt{\beta_{k+1}^2e^2-4ce} \\ \beta_{k+1}=\dfrac{p}{2}\left(1-\cosh\left(\dfrac{1}{2}\cosh^{-1}\left(\dfrac{16ce^2p-4c^4-3e^3p^3}{e^3p^3}\right)\right)+i\pi k\right), \ k=0,1 \\ p^2=\dfrac{8ec}{3e^2}+\dfrac{4c}{3e^2}\sqrt{3c^2d+10e^2}\cosh\left(\dfrac{1}{3}\cosh^{-1}\left(\dfrac{288c^2de+27c^5+448e^3}{16(3c^2d+10e^2)^{3/2}}\right)+\dfrac{i2\pi k}{3}\right). \end{cases}$$

$$\sqrt{a\pm ib}=\dfrac{\sqrt{2}}{2}\sqrt{\sqrt{a^2+b^2}+a}\pm i\dfrac{\sqrt{2}}{2}\sqrt{\sqrt{a^2+b^2}-a}$$

$$(a\pm ib)^{1/4}=\dfrac{1}{2}(a^2+b^2)^{1/8}\left[\sqrt{2+\sqrt{2+\dfrac{2}{\sqrt{1+\left(\dfrac{b}{a}\right)^2}}}}\pm i\sqrt{2-\sqrt{2+\dfrac{2}{\sqrt{1+\left(\dfrac{b}{a}\right)^2}}}}\right]$$

$$i=\sqrt{-1}$$

$$f=+-\times\div$$

$$e^{i\pi}=-1$$

TUE VU

SMS – Series Math Study
Copyright ©2024 Tue Vu.

ISBN 979-8-218-97352-0
Website: www.seriesmathstudy.com

ISBN 979-8-218-97352-0

9 798218 973520 >

Preface

The publication focuses on two elementary mathematical subjects: the exploration of nested radicals and algebraic identities, and the art of solving polynomial equations using of trigonometric and hyperbolic trigonometric functions. The first subject unveils beautiful nested radical identities in algebra, along with essential known formulas in mathematics. Meanwhile, the second subject unveils an innovative approach to solving polynomial equations, encompassing quadratic to quartic and other higher order solvable equations, by utilizing trigonometric and hyperbolic trigonometric functions. We have chosen subjects that are relevant to elementary algebra in mathematics. Our aim is to refresh a small facet of these subjects by adopting an unconventional yet intentionally creative approach.

Our preface reveals that we are not professional mathematicians but rather passionate math enthusiasts. By self-publishing this book, we recognize that it represents merely a small fraction of our perception of mathematics, given our limited expertise. Despite the modest content, our enthusiasm effectively communicates the core mathematical results of these subjects in concise and clear forms to the readers. The use of terms like *innovation*, *discovery* or *exploration* in this book refers to the author, and while the results may not be new to readers, they hold personal significance for us. We anticipate that this e-book will serve as a memorable resource for mathematical exchange. The e-book is available at a very low cost. Please note that this pricing is specific to the e-book format and does not apply to the paperback version, due to the additional expenses associated with printing and shipping for those interested in obtaining a physical copy. For instance, we have set the e-book's price at a level covering only the expenses of self-publishing on Amazon. A few words from the author: Grateful for the encouragement from my parents and three daughters, Thao Vu, Thanh-Tu Vu, and Mai-Tran Vu, who are currently pursuing their education. To the community, thank you for everything.

While this book may exhibit variations in writing style influenced by cultural factors, as well as potential errors such as incorrect results, typos, and other imperfections, we sincerely welcome constructive feedback and contributions from readers to improve the overall quality. Your valuable input is greatly appreciated. Our contact emails are sms@seriesmathstudy.com and tuevu2003@gmail.com. If you wish to support our book(s) and efforts, we gratefully accept donations without refunds in any form through the provided PayPal link, https://www.paypal.com/donate/?hosted_button_id=87LZ956NGUMQQ.

Table of Contents

Chapter 1

Nested Radicals and Algebraic Identities

This chapter explores into nested radical formulas rooted in trigonometry, establishing some algebraic identities involving nested radicals. Additionally, the exploration extends beyond trigonometry to include various other forms of nested radicals.

Section 1. Nested Radicals Generated by Composition of Trigonometric Functions with Inverse Trigonometric Functions, and Hyperbolic Trigonometric Functions with Inverse Hyperbolic Functions

We introduce a creative approach to derive nested radical expressions by composing various trigonometric and hyperbolic trigonometric functions with inverse functions. Our starting point involves utilizing the cosine double angle formula, a key component in constructing and revealing nested radical patterns. We elucidate the process of creating nested radical addition and subtraction identities by employing the definitions of these composite functions. This represents a significant advancement in the realm of nested radical expressions, particularly in the context of trigonometric functions, enabling the rapid and precise computation of specific values. As an illustrative example, we demonstrate the accurate evaluation of expressions without using calculator such as cos(π/12), cos(π/24) and cos(π/48) in closed form.

Section 1-1. Nested Radical Generated by Composition cos[(1/2ⁿ)cos⁻¹(x)]

We use cosine double angle formula to construct a set of finite nested radical expressions in relation to the composite of the cosine with its inverse in the form:

$$\cos\left(\frac{1}{2^n}\cos^{-1}(x)\right), \quad -1 \le x \le 1, \quad n \in \mathbb{N} = \{1,2,3,\dots\} \tag{1.1}$$

where cos⁻¹(x) is the inverse function of cosine. The expression (1.1) has domain [-1,

1] and range $(-\infty, \infty)$.

1. Half Angle Cosine Formula

The half angle cosine formula is expressed as

$$\cos\left(\frac{1}{2}\cos^{-1}(x)\right).$$

To derive this formula, we use the cosine double angle formula, which can be expressed in relation to $\cos(x)$ as

$$\cos 2x = 2\cos^2 x - 1$$

Equivalently, it can be expressed in the form,

$$\cos x = \pm\frac{1}{2}\sqrt{(2\cos 2x + 2)}.$$

Substituting x by x/2 gives

$$\cos\frac{x}{2} = \pm\frac{1}{2}\sqrt{2 + 2\cos x}. \tag{1.2}$$

Continuing by replacing x with $\cos^{-1}(x)$, for $-1 \leq x \leq 1$, we obtain

$$\cos\left(\frac{1}{2}\cos^{-1}(x)\right) = \frac{1}{2}\sqrt{2+2x} = \frac{1}{\sqrt{2}}\sqrt{1+x}. \tag{1.3}$$

The negative sign is omitted due to the restrictions of the inverse cosine function's domain and range, which are $[-1, 1]$ and $[0, \pi]$, respectively. Moving forward, the negative sign will be excluded. Noting that the expression on left-hand side of (1.3) also follows from (1.1) for n = 1.

2. Quarter Angle Cosine Formula

The quarter angle cosine formula that we want to derive has in the form:

$$\cos\left(\frac{1}{4}\cos^{-1}(x)\right).$$

Upon substituting x by x/2 into (1.2), gives

$$\cos\left(\frac{x}{4}\right) = \frac{1}{2}\sqrt{2 + 2\cos\left(\frac{x}{2}\right)}$$

2

$$=\frac{1}{2}\sqrt{2+\sqrt{2+2\cos(x)}}.\tag{1.4}$$

Substituting x with $\cos^{-1}(x)$, for $-1\leq x\leq1$, into (1.4) gives the formula,

$$\cos\left(\frac{1}{4}\cos^{-1}(x)\right)=\frac{1}{2}\sqrt{2+\sqrt{2+2x}}\ \ \text{for}-1\leq x\leq1.\tag{1.5}$$

Proof of Domain Determination for (1.5)
a. On the right-hand side of (1.5), we find that:

$$\begin{cases}2+2x\geq0\\2+\sqrt{2+2x}\geq0\end{cases}\Rightarrow\begin{cases}x\geq-1\\x\geq-1\end{cases}\Rightarrow x\geq-1.\tag{1.5a}$$

b. On the left-hand side of (1.5), we determine: $1\leq x\leq1$.

By combining the results from Part a and Part b, we conclude that the domain of (1.5) is the interval [-1, 1].

3. Other Cosine Identities for Higher Roots
Following the established procedure, we derive a set of nested radical identities for the scaling factor $1/2^n$ at 8th, 16th, and 2^{nth} which follows from setting n = 3, 4,..., n in (1.1), as illustrated below:

$$\cos\left(\frac{1}{8}\cos^{-1}(x)\right)=\frac{1}{2}\sqrt{2+\sqrt{2+\sqrt{2+2x}}},\ \ -1\leq x\leq1,\tag{1.6}$$

$$\cos\left(\frac{1}{16}\cos^{-1}(x)\right)=\frac{1}{2}\sqrt{2+\sqrt{2+\sqrt{2+\sqrt{2+2x}}}},\ \ -1\leq x\leq1,\tag{1.7}$$

$$\cos\left(\frac{1}{32}\cos^{-1}(x)\right)=\frac{1}{2}\sqrt{2+\sqrt{2+\sqrt{2+\sqrt{2+\sqrt{2+2x}}}}},\ \ -1\leq x\leq1,\tag{1.8}$$

$$\cos\left(\frac{1}{64}\cos^{-1}(x)\right)=\frac{1}{2}\sqrt{2+\sqrt{2+\sqrt{2+\sqrt{2+\sqrt{2+\sqrt{2+2x}}}}}},\ \ -1\leq x\leq1,\tag{1.9}$$

$$\tag{1.10}$$

...

$$\cos\left(\frac{1}{2^n}\cos^{-1}(x)\right)=\frac{1}{2}\cdot\sqrt{2+...+\sqrt{2+\sqrt{2+\sqrt{2+\sqrt{2+2x}}}}},\ \ -1\leq x\leq1,\ n\in\mathbb{N},\tag{1.10a}$$

3

or displaying (1.10a) in "another form style",

$$\cos\left(\frac{1}{2^n}\cos^{-1}x\right)=\frac{1}{2}\cdot\sqrt{\sqrt{\sqrt{\sqrt{\sqrt{2x+2}+2}+2}+2}+\ldots+2}, \qquad -1\le x\le 1, \ n\in\mathbb{N}. \qquad (1.10\,b)$$

The expression (1.10a) or (1.10b) consists of n nested square roots (or nested radical square roots). The initial layer begins with the square root of 2x + 2, and each successive layer builds upon the previous one by containing 2.

Example. Calculate cos(π/12), cos(π/24) and cos(π/48) manually, without relying on a calculator.

Solution. We use the fact that cos⁻¹(1/2) = π/3, as a starting point. By substituting x = ½ into the formulas (1.5), (1.6), and (1.7), we can efficiently determine the values of cos(π/12), cos(π/24) and cos(π/48), resulting in the following:

$$\cos\left(\frac{\pi}{12}\right)=\frac{1}{2}\sqrt{2+\sqrt{3}},$$

$$\cos\left(\frac{\pi}{24}\right)=\frac{1}{2}\sqrt{2+\sqrt{2+\sqrt{3}}},$$

and

$$\cos\left(\frac{\pi}{48}\right)=\frac{1}{2}\sqrt{2+\sqrt{2+\sqrt{2+\sqrt{3}}}}.$$

Section 1-2. Nested Radical Generated by Composition cosh[(1/2ⁿ)cosh⁻¹(x)]

In this section, we develop a set of the nested radical expressions in relation to the hyperbolic cosine and its inverse [1] with the structure:

$$\cosh\left(\frac{1}{2^n}\cosh^{-1}(x)\right) \text{ for } x\ge 1 \text{ and } n=\{1,2,3,\ldots\}. \qquad (1.2.1)$$

where cosh⁻¹(x) is the inverse function of cosh(x).

1. Half Angle Hyperbolic Cosine Formula

The half angle hyperbolic cosine formula is expressed as

$$\cosh\left(\frac{1}{2}\cosh^{-1}(x)\right), \quad x \geq 1,$$

which is the form (1.2.1) when n = 1. To derive this formula, we use the identity $\cos(ix) = \cosh(x)$. By substituting x with ix into (1.2), we deduce that

$$\cosh\left(\frac{x}{2}\right) = \frac{1}{2}\sqrt{2 + 2\cosh(x)}. \tag{1.2.2}$$

Continue substituting x by $\cosh^{-1}(x)$ into (1.2.2) yields,

$$\cosh\left(\frac{1}{2}\cosh^{-1}(x)\right) = \frac{1}{2}\sqrt{2 + 2x}, \quad x \geq 1. \tag{1.2.2 a}$$

The expression $(1/2)\sqrt{2 + 2x}$ in (1.2.2a) is intentionally left in its non-simplified form to preserve and analyze the underlying patterns within the formula.

2. Quarter Angle Hyperbolic Cosine Formula

The quarter angle hyperbolic cosine formula is expressed in the form,

$$\cosh\left(\frac{1}{4}\cosh^{-1}(x)\right).$$

To construct this formula, we replace x with x/2 into (1.2.2), which gives

$$\cosh\left(\frac{x}{4}\right) = \frac{1}{2}\sqrt{2 + 2\cosh\left(\frac{x}{2}\right)}$$

$$= \frac{1}{2}\sqrt{2 + \sqrt{2 + 2\cosh(x)}}.$$

Substituting x by $\cosh^{-1}(x)$ gives the final result

$$\cosh\left(\frac{1}{4}\cosh^{-1}(x)\right) = \frac{1}{2}\sqrt{2 + \sqrt{2 + 2x}}, \quad x \geq 1. \tag{1.2.3}$$

Notice that the expression on left-hand hand of (1.2.3) is also the value of form (1.2.1) when n = 2.

3. Other Hyperbolic Cosine Identities for Higher Roots

Continuing in a similar manner as described in parts 1 and 2, we derive a set of the nested radical identities for the scaling factor $1/2^n$ at 8th, 16th,..., 2^{nth}, which follows from setting n = 3, 4,..., n in (1.2.1), as illustrated below:

$$\cosh\left(\frac{1}{8}\cosh^{-1}(x)\right)=\frac{1}{2}\sqrt{2+\sqrt{2+\sqrt{2+2x}}},\quad x\geq 1,\tag{1.2.4}$$

$$\cosh\left(\frac{1}{16}\cosh^{-1}(x)\right)=\frac{1}{2}\sqrt{2+\sqrt{2+\sqrt{2+\sqrt{2+2x}}}},\quad x\geq 1,\tag{1.2.5}$$

$$\cosh\left(\frac{1}{32}\cosh^{-1}(x)\right)=\frac{1}{2}\sqrt{2+\sqrt{2+\sqrt{2+\sqrt{2+\sqrt{2+2x}}}}},\quad x\geq 1,\tag{1.2.6}$$

$$\cosh\left(\frac{1}{64}\cosh^{-1}(x)\right)=\frac{1}{2}\sqrt{2+\sqrt{2+\sqrt{2+\sqrt{2+\sqrt{2+\sqrt{2+2x}}}}}},\quad x\geq 1,\tag{1.2.7}$$

$$\cosh\left(\frac{1}{128}\cosh^{-1}(x)\right)=\frac{1}{2}\sqrt{2+\sqrt{2+\sqrt{2+\sqrt{2+\sqrt{2+\sqrt{2+\sqrt{2+2x}}}}}}},\quad x\geq 1,\tag{1.2.8}$$

...

$$\cosh\left(\frac{1}{2^n}\cosh^{-1}x\right)=\frac{1}{2}\cdot\sqrt{2+...\sqrt{2+\sqrt{2+\sqrt{2+\sqrt{2+\sqrt{2+\sqrt{2+2x}}}}}}},\quad x\geq 1,\ n\in\mathbb{N}.\tag{1.2.9}$$

Note: Formula (1.2.9) comprises n layers of nested radical square roots. The initial layer begins with the square root of (2x + 2), and each successive layer builds upon the previous one by containing 2.

Section 1-3. Establish Nested Radical Addition Identities

In this section, we use the definition of hyperbolic cosine and its inverse, along with the archived results from Section 1-2, to establish algebraic identities for nested radical addition.

The definition of the hyperbolic cosine function is defined as follows:

$$\cosh(x)=\frac{1}{2}\left(e^x+e^{-x}\right),\quad x\in\mathbb{R}.\tag{1.3.1}$$

Likewise, the inverse hyperbolic cosine function is defined as

$$\cosh^{-1}(x)=\ln\left(x+\sqrt{x^2-1}\right)\text{ for }x\geq 1.\tag{1.3.2}$$

6

For real n, we have

$$\cosh\left(\frac{1}{n}\cosh^{-1}(x)\right)=\cosh\left(\frac{1}{n}\ln\left(x+\sqrt{x^2-1}\right)\right) \tag{1.3.3}$$

$$=\cosh\left(\ln\left(x+\sqrt{x^2-1}\right)^{1/n}\right)$$

$$=\frac{1}{2}\left(x+\sqrt{x^2-1}\right)^{1/n}+\frac{1}{2}\left(x+\sqrt{x^2-1}\right)^{-1/n}$$

$$=\frac{1}{2}\left(x+\sqrt{x^2-1}\right)^{1/n}+\frac{1}{2}\left(x-\sqrt{x^2-1}\right)^{1/n}. \tag{1.3.3a}$$

By replacing n with 2^n for $n\in\mathbb{N}$, we obtain the hyperbolic cosine and its inverse in the form:

$$\cosh\left(\frac{1}{2^n}\cosh^{-1}(x)\right)=\frac{1}{2}\left(x+\sqrt{x^2-1}\right)^{1/2^n}+\frac{1}{2}\left(x-\sqrt{x^2-1}\right)^{1/2^n},n=1,2,3,... \tag{1.3.3b}$$

1. Nested Radical Addition Identities

By employing the outcomes derived from (1.2.2a) through (1.2.9) in conjunction with formula (1.3.3b) for n = 1, 2, 3,..., we unveil the elegant identities presented below:

$$\sqrt{x+\sqrt{x^2-1}}+\sqrt{x-\sqrt{x^2-1}}=\sqrt{2+2x}, \tag{1.3.4}$$

$$\sqrt[4]{x+\sqrt{x^2-1}}+\sqrt[4]{x-\sqrt{x^2-1}}=\sqrt{2+\sqrt{2+2x}}, \tag{1.3.5}$$

$$\sqrt[8]{x+\sqrt{x^2-1}}+\sqrt[8]{x-\sqrt{x^2-1}}=\sqrt{2+\sqrt{2+\sqrt{2+2x}}}, \tag{1.3.6}$$

$$\sqrt[16]{x+\sqrt{x^2-1}}+\sqrt[16]{x-\sqrt{x^2-1}}=\sqrt{2+\sqrt{2+\sqrt{2+\sqrt{2+2x}}}}, \tag{1.3.7}$$

$$\sqrt[32]{x+\sqrt{x^2-1}}+\sqrt[32]{x-\sqrt{x^2-1}}=\sqrt{2+\sqrt{2+\sqrt{2+\sqrt{2+\sqrt{2+2x}}}}}, \tag{1.3.8}$$

$$\sqrt[64]{x+\sqrt{x^2-1}}+\sqrt[64]{x-\sqrt{x^2-1}}=\sqrt{2+\sqrt{2+\sqrt{2+\sqrt{2+\sqrt{2+\sqrt{2+2x}}}}}}, \tag{1.3.9}$$

$$\sqrt[128]{x+\sqrt{x^2-1}}+\sqrt[128]{x-\sqrt{x^2-1}}=\sqrt{2+\sqrt{2+\sqrt{2+\sqrt{2+\sqrt{2+\sqrt{2+\sqrt{2+2x}}}}}}}, \tag{1.3.10}$$

...

$$\sqrt[2^n]{x+\sqrt{x^2-1}}+\sqrt[2^n]{x-\sqrt{x^2-1}}=\sqrt{2+...\sqrt{2+\sqrt{2+\sqrt{2+\sqrt{2+2x}}}}}, \tag{1.3.11}$$

where x ≥ 1 for each of these identities spanning from (1.3.4) to (1.3.11). Formula (1.3.11) comprises n layers of nested radical square roots. The initial layer begins with the square root of 2x + 2, and each successive layer builds upon the previous one by containing 2.

2. Alternative Form of Nested Radical Addition Identities

We look for an alternative form of part 1 so that it shows more symmetry in algebraic structure. We begin with (1.3.4) as follows:

- Substituting x by $\sqrt{x}$ into (1.3.4) gives

$$\sqrt{\sqrt{x}+\sqrt{x-1}}+\sqrt{\sqrt{x}-\sqrt{x-1}}=\sqrt{2+2\sqrt{x}}. \qquad (1.3.12)$$

- Let a and b are arbitrary, by replacing x with a/(2b) + 1/2 into (1.3.12), which gives

$$\sqrt{\sqrt{\frac{a}{2b}+\frac{1}{2}}+\sqrt{\frac{a}{2b}-\frac{1}{2}}}+\sqrt{\sqrt{\frac{a}{2b}+\frac{1}{2}}-\sqrt{\frac{a}{2b}-\frac{1}{2}}}=\sqrt{2+2\sqrt{\frac{a}{2b}+\frac{1}{2}}}. \qquad (1.3.13)$$

After rearranging for a common denominator 2b, and simplifying the shared denominator $\sqrt{\sqrt{2b}}$ or $\sqrt[4]{2b}$ on both sides of (1.3.13), we obtain a graceful nested radical identity or name it as a nested radical addition identity, namely

$$\sqrt{\sqrt{a+b}+\sqrt{a-b}}+\sqrt{\sqrt{a+b}-\sqrt{a-b}}=\sqrt{2\left(\sqrt{2b}+\sqrt{a+b}\right)} \qquad (1.3.14)$$

In the same manner, employing the identical approach yields other notably elegant nested radical addition identities for values of n corresponding to powers of 2, 4, 8,..., as illustrated below:

$$\sqrt[4]{\sqrt{a+b}+\sqrt{a-b}}+\sqrt[4]{\sqrt{a+b}-\sqrt{a-b}}=\sqrt{2\sqrt[4]{2b}+\sqrt{2\left(\sqrt{2b}+\sqrt{a+b}\right)}} \qquad (1.3.15)$$

$$\sqrt[8]{\sqrt{a+b}+\sqrt{a-b}}+\sqrt[8]{\sqrt{a+b}-\sqrt{a-b}}=\sqrt{2\sqrt[8]{2b}+\sqrt{2\sqrt[4]{2b}+\sqrt{2\left(\sqrt{2b}+\sqrt{a+b}\right)}}} \qquad (1.3.16)$$

$$\sqrt[16]{\sqrt{a+b}+\sqrt{a-b}}+\sqrt[16]{\sqrt{a+b}-\sqrt{a-b}}=\sqrt{2\sqrt[16]{2b}+\sqrt{2\sqrt[8]{2b}+\sqrt{2\sqrt[4]{2b}+\sqrt{2\left(\sqrt{2b}+\sqrt{a+b}\right)}}}} \qquad (1.3.17)$$

$$\sqrt[32]{\sqrt{a+b}+\sqrt{a-b}}+\sqrt[32]{\sqrt{a+b}-\sqrt{a-b}}=\sqrt{2\sqrt[32]{2b}+\sqrt{2\sqrt[16]{2b}+\sqrt{2\sqrt[8]{2b}+\sqrt{2\sqrt[4]{2b}+\sqrt{2\left(\sqrt{2b}+\sqrt{a+b}\right)}}}}} \qquad (1.3.18)$$

$$\sqrt[64]{\sqrt{a+b}+\sqrt{a-b}}+\sqrt[64]{\sqrt{a+b}-\sqrt{a-b}}=\sqrt{2\sqrt[64]{2b}+\sqrt{2\sqrt[32]{2b}+\sqrt{2\sqrt[16]{2b}+\sqrt{2\sqrt[8]{2b}+\sqrt{2\sqrt[4]{2b}+\sqrt{2\left(\sqrt{2b}+\sqrt{a+b}\right)}}}}}} \qquad (1.3.19)$$

$$\sqrt[128]{\sqrt{a+b}+\sqrt{a-b}}+\sqrt[128]{\sqrt{a+b}-\sqrt{a-b}}=\sqrt{2\sqrt[128]{2b}+\sqrt{2\sqrt[64]{2b}+\sqrt{2\sqrt[32]{2b}+\sqrt{2\sqrt[16]{2b}+\sqrt{2\sqrt[8]{2b}+\sqrt{2\sqrt[4]{2b}+\sqrt{2\left(\sqrt{2b}+\sqrt{a+b}\right)}}}}}}} \qquad (1.3.20)$$

...

$$\sqrt[2^n]{\sqrt{a+b}+\sqrt{a-b}}+\sqrt[2^n]{\sqrt{a+b}-\sqrt{a-b}}=\sqrt{2\sqrt[2^n]{2b}+\ldots+\sqrt{2\sqrt[4]{2b}+\sqrt{2\left(\sqrt{2b}+\sqrt{a+b}\right)}}},\, a,b\in\mathbb{R},\, n\in N \qquad (1.3.21)$$

Note: The expression (1.3.21) consists of n nested square roots, where each successive layer builds upon the previous one by containing $2\sqrt{2b}$.

3. Second Alternative Form of Radical Square Addition Identities

The second alternative form of Radical Square Addition Identities is another form of part (2) which can reformulated to a simpler form such that the square roots of (a+b) and (a-b) of each identity spanning from (1.3.4) to (1.3.14) are removed. Indeed, let A and B be real numbers, and by setting: $A=\sqrt{a+b}$ and $B=\sqrt{a-b}$, we derive upon solving for a and b in terms of A and B, which yields

$$a=\frac{A^2+B^2}{2}$$

and

$$b=\frac{A^2-B^2}{2}.$$

Substituting a and b into the identities found between (1.3.14) and (1.3.21), we uncover the subsequent intriguingly and remarkably nested radical addition identities:

$$\sqrt{A+B}+\sqrt{A-B}=\sqrt{2A+2\sqrt{A^2-B^2}}, \qquad (1.3.22)$$

$$\sqrt[4]{A+B}+\sqrt[4]{A-B}=\sqrt{2\sqrt[4]{A^2-B^2}+\sqrt{2A+2\sqrt{A^2-B^2}}}, \qquad (1.3.23)$$

$$\sqrt[8]{A+B}+\sqrt[8]{A-B}=\sqrt{2\sqrt[8]{A^2-B^2}+\sqrt{2\sqrt[4]{A^2-B^2}+\sqrt{2A+2\sqrt{A^2-B^2}}}}, \qquad (1.3.24)$$

$$\sqrt[16]{A+B}+\sqrt[16]{A-B}=\sqrt{2\sqrt[16]{A^2-B^2}+\sqrt{2\sqrt[8]{A^2-B^2}+\sqrt{2\sqrt[4]{A^2-B^2}+\sqrt{2A+2\sqrt{A^2-B^2}}}}},\qquad (1.3.25)$$

$$\sqrt[32]{A+B}+\sqrt[32]{A-B}=\sqrt{2\sqrt[32]{A^2-B^2}+\sqrt{2\sqrt[16]{A^2-B^2}+\sqrt{2\sqrt[8]{A^2-B^2}+\sqrt{2\sqrt[4]{A^2-B^2}+\sqrt{2A+2\sqrt{A^2-B^2}}}}}},\quad (1.3.26)$$

$$\sqrt[64]{A+B}+\sqrt[64]{A-B}=\sqrt{2\sqrt[64]{A^2-B^2}+\sqrt{2\sqrt[32]{A^2-B^2}+\sqrt{2\sqrt[16]{A^2-B^2}+\sqrt{2\sqrt[8]{A^2-B^2}+\sqrt{2\sqrt[4]{A^2-B^2}+\sqrt{2A+2\sqrt{A^2-B^2}}}}}}},\qquad (1.3.27)$$

...

$$\sqrt[2^n]{A+B}+\sqrt[2^n]{A-B}=\sqrt{2\sqrt[2^n]{A^2-B^2}+...+\sqrt{2\sqrt[2^{n-1}]{A^2-B^2}+\sqrt{2A+2\sqrt{A^2-B^2}}}},\ A,B\in\mathbb{R},\ n\in N.\quad (1.3.28)$$

Formula (1.3.28) comprises n layers of nested square roots. Each successive layer builds upon the previous one by including the difference between the squares of A and B. Notice that A and B can be arbitrary numbers.

Thus, we have effectively demonstrated a straightforward algebraic method for deriving elegant algebraic identities pertaining to nested radicals of 2, 4, 8, and so forth. However, in our exploration, we haven't found equivalent elegant 2-term identities for nested radicals with odd numbers. The presence of such identities, devoid of the need for complex numbers within them, is still unknown and uncertain at this point. Interestingly, if we begin with the cosine of 3rd angle in Section 1-2, it leads us to solve a cubic equation. Surprisingly, the outcome aligns with the form of formula (1.3.3a) in which its radical expression involves the complex unit i.

There exists a mathematical identity for the cubic radical such that
$$\sqrt[3]{A+B}+\sqrt[3]{A-B}=\sqrt[3]{\text{expressed without the complex unit i}}\ ?$$

Section 1-4. Nested Radical Generated by Composition sin[(1/2ⁿ)cos⁻¹(x)]

Within this section, we construct nested radical expressions involving sine function and inverse cosine function in the form:

$$\sin\left(\frac{1}{2^n}\cos^{-1}(x)\right),\, -1\le x\le 1,\, n\in \mathbb{N}=\{1,2,3,...\}. \qquad (1.4.1)$$

1. Half Angle Sine Formula

The cosine double angle formula can be expressed in the form:

$$\cos 2x = 1 - 2\sin^2 x$$

Alternatively, it can also be stated as

$$\sin(x) = \pm\frac{1}{2}\sqrt{2 - 2\cos(2x)}.$$

Then replacing x with x/2 gives

$$\sin\left(\frac{x}{2}\right) = \pm\frac{1}{2}\sqrt{2 - 2\cos(x)} \qquad (1.4.2)$$

By replacing x with $\cos^{-1}(x)$ into (1.4.2), we obtain

$$\sin\left(\frac{1}{2}\cos^{-1}(x)\right) = \frac{1}{2}\sqrt{2 - 2x}, \quad -1 \leq x \leq 1. \qquad (1.4.3)$$

The negative sign is not included in (1.4.3) due to the restrictions on the domain and range of the inverse cosine function. The expression on left-hand side of (1.4.3) corresponds to form (1.4.1) when n = 1.

2. Quarter Angle Sine Formula
We derive the quarter angle sine formula in the form:

$$\sin\left(\frac{1}{4}\cos^{-1}(x)\right).$$

By replacing x with x/2 into (1.4.2), we have

$$\sin\left(\frac{x}{4}\right) = \pm\frac{1}{2}\sqrt{2 - 2\cos\left(\frac{x}{2}\right)}$$

$$= \pm\frac{1}{2}\sqrt{2 - \sqrt{2 + 2\cos(x)}} \qquad (1.4.4)$$

Replacing x with $\cos^{-1}(x)$ into (1.4.4), for $-1 \leq x \leq 1$, gives

$$\sin\left(\frac{1}{4}\cos^{-1}(x)\right) = \frac{1}{2}\sqrt{2 - \sqrt{2 + 2x}}. \qquad (1.4.5)$$

3. Other Sine Identities for Higher Roots
In the similar manner, we derive other cases of (1.4.1) for n = 3, 4,...n, with $-1 \leq x \leq 1$ as indicated below

$$\sin\left(\frac{1}{8}\cos^{-1}(x)\right)=\frac{1}{2}\sqrt{2-\sqrt{2+\sqrt{2+2x}}},\tag{1.4.6}$$

$$\sin\left(\frac{1}{16}\cos^{-1}(x)\right)=\frac{1}{2}\sqrt{2-\sqrt{2+\sqrt{2+\sqrt{2+2x}}}},\tag{1.4.7}$$

$$\sin\left(\frac{1}{32}\cos^{-1}(x)\right)=\frac{1}{2}\sqrt{2-\sqrt{2+\sqrt{2+\sqrt{2+\sqrt{2+2x}}}}},\tag{1.4.8}$$

$$\sin\left(\frac{1}{64}\cos^{-1}(x)\right)=\frac{1}{2}\sqrt{2-\sqrt{2+\sqrt{2+\sqrt{2+\sqrt{2+\sqrt{2+2x}}}}}},\tag{1.4.9}$$

$$\sin\left(\frac{1}{128}\cos^{-1}(x)\right)=\frac{1}{2}\sqrt{2-\sqrt{2+\sqrt{2+\sqrt{2+\sqrt{2+\sqrt{2+2x}}}}}},\tag{1.4.10}$$

...

$$\sin\left(\frac{1}{2^n}\cos^{-1}x\right)=\frac{1}{2}\cdot\sqrt{2-\sqrt{2...+\sqrt{2+\sqrt{2+\sqrt{2+\sqrt{2+\sqrt{2+2x}}}}}}},\quad -1\le x\le 1.\tag{1.4.11}$$

Note that another approach is that the identities uncovered in the Section 1-1 can be promptly employed to deduce the aforementioned outcomes. This can be achieved by utilizing the trigonometric identity $\sin^2 x + \cos^2 x = 1$, for instance, we choose formula (1.6),

$$\cos\left(\frac{1}{8}\cos^{-1}(x)\right)=\frac{1}{2}\sqrt{2+\sqrt{2+\sqrt{2+2x}}}.$$

Then take square of both its sides, gives

$$\cos^2\left(\frac{1}{8}\cos^{-1}(x)\right)=\frac{1}{4}\left(2+\sqrt{2+\sqrt{2+2x}}\right)$$

$$\Leftrightarrow \sin^2\left(\frac{1}{8}\cos^{-1}(x)\right)=1-\frac{1}{4}\left(2+\sqrt{2+\sqrt{2+2x}}\right)$$

$$\Rightarrow \sin\left(\frac{1}{8}\cos^{-1}(x)\right)=\frac{1}{2}\sqrt{2-\sqrt{2+\sqrt{2+2x}}},$$

which was previously demonstrated in (1.4.6).

Example.

Find the exact values of sin(π/48) and sin(π/96).

We use the fact that $\cos^{-1}(1/2) = \pi/3$, as a starting point. By substituting x = ½ into formulas (1.4.7) and (1.4.8), we can efficiently determine the values of sin(π/48) and sin(π/96), resulting in the following:

$$\sin\left(\frac{\pi}{48}\right) = \frac{1}{2}\sqrt{2-\sqrt{2+\sqrt{2+\sqrt{3}}}}$$

$$\sin\left(\frac{\pi}{96}\right) = \frac{1}{2}\sqrt{2-\sqrt{2+\sqrt{2+\sqrt{2+\sqrt{3}}}}}.$$

Section 1-5. Nested Radical Generated by Composition sinh[(1/2ⁿ)cosh⁻¹(x)]

We employ the identities revealed in Section 1-4 to construct nested radical expressions that in relation to the composite of the hyperbolic sine function with inverse hyperbolic cosine function [2] in the form:

$$\sinh\left(\frac{1}{2^n}\cosh^{-1}(x)\right), x \geq 1, n \in N = \{1,2,3,...\} \tag{1.5.1}$$

1. Half Angle Hyperbolic Sine

The half angle hyperbolic sine function involving inverse hyperbolic cosine function is a special case of (1.5.1) when n = 1.

To derive it, we do in the following steps:

a. Multiplying both sides of (1.4.2) by i, we have

$$i\sin\left(\frac{1}{2}\cos^{-1}(x)\right) = \frac{i}{2}\sqrt{2-2x}$$

$$\Rightarrow \sinh\left(\frac{i}{2}\cos^{-1}(x)\right) = \frac{1}{2}\sqrt{2x-2}$$

b. Applying the identity icos⁻¹(x) = cosh(x), we obtain the half angle hyperbolic sine,

$$\sinh\left(\frac{1}{2}\cosh^{-1}(x)\right) = \frac{1}{2}\sqrt{2x-2}, x \geq 1. \tag{1.5.2}$$

2. Quarter Angle Hyperbolic Sine

The quarter angle hyperbolic sine is a special case of (1.5.2) when n = 2. To derive it, we do in the following steps:

a. Replacing x with cosh(x) into (1.5.2), then substituting x with x/2 we have

$$\sinh\left(\frac{x}{4}\right)=\frac{1}{2}\sqrt{2\cosh\left(\frac{x}{2}\right)-2}.$$

b. Continue replacing x with $\cosh^{-1}(x)$ for x ≥ 1 gives

$$\sinh\left(\frac{1}{4}\cosh^{-1}(x)\right)=\frac{1}{2}\sqrt{2\cosh\left(\frac{1}{2}\cosh^{-1}(x)\right)-2}$$

c. Applying (1.2.2a) we obtain the quarter angle hyperbolic sine, namely

$$\sinh\left(\frac{1}{4}\cosh^{-1}(x)\right)=\frac{1}{2}\sqrt{\sqrt{2+2x}-2},\quad x\geq1. \tag{1.5.3}$$

3. Other Hyperbolic Sine Identities for Higher Roots

Following the established procedure, we proceed to deduce the subsequent identities of (1.5.1) when n = 3, 4, 5, etc., as presented below:

$$\sinh\left(\frac{1}{8}\cosh^{-1}(x)\right)=\frac{1}{2}\sqrt{\sqrt{2+\sqrt{2+2x}}-2},\quad x\geq1, \tag{1.5.4}$$

$$\sinh\left(\frac{1}{16}\cosh^{-1}(x)\right)=\frac{1}{2}\sqrt{\sqrt{2+\sqrt{2+\sqrt{2+2x}}}-2},\quad x\geq1, \tag{1.5.5}$$

$$\sinh\left(\frac{1}{32}\cosh^{-1}(x)\right)=\frac{1}{2}\sqrt{\sqrt{2+\sqrt{2+\sqrt{2+\sqrt{2+2x}}}}-2},\quad x\geq1, \tag{1.5.6}$$

$$\sinh\left(\frac{1}{64}\cosh^{-1}(x)\right)=\frac{1}{2}\sqrt{\sqrt{2+\sqrt{2+\sqrt{2+\sqrt{2+\sqrt{2+2x}}}}}-2},\quad x\geq1, \tag{1.5.7}$$

$$\sinh\left(\frac{1}{128}\cosh^{-1}(x)\right)=\frac{1}{2}\sqrt{\sqrt{2+\sqrt{2+\sqrt{2+\sqrt{2+\sqrt{2+\sqrt{2+2x}}}}}}-2},\quad x\geq1, \tag{1.5.8}$$

...

14

$$\sinh\left(\frac{1}{2^n}\cosh^{-1}(x)\right)=\frac{1}{2}\sqrt{2+\dots\sqrt{2+\sqrt{2+\sqrt{2+\sqrt{2+\sqrt{2+2x-2}}}}}}, \ x\geq 1.$$ (1.5.9)

(n nested square roots)

Note: Formula (1.5.9) comprises n layers of nested square roots. The initial layer begins with the square root of 2x - 2, and each successive layer builds upon the previous one by including 2.

Section 1-6. Establish Nested Radical Subtraction Identities

Following a similar approach to the Section 1-3, we establish a set of algebraic identities referred to as nested radical subtraction identities. The identities are derived through the utilization of the definitions of hyperbolic sine and inverse hyperbolic cosine functions, along with the outcomes obtained in Section 1-4. We derive the general formula for $\sinh((1/2^n)\cosh^{-1}(x))$ as follows:

The definition of the hyperbolic sine function is defined as

$$\sinh(x)=\frac{1}{2}\left(e^x-e^{-x}\right), \ x\in\mathbb{R}.$$ (1.6.1)

Likewise, the inverse hyperbolic cosine function is defined as

$$\cosh^{-1}(x)=\ln\left(x+\sqrt{x^2-1}\right), \ x\geq 1.$$ (1.6.2)

For real n, we have

$$\sinh\left(\frac{1}{n}\cosh^{-1}(x)\right)=\sinh\left(\frac{1}{n}\ln\left(x+\sqrt{x^2-1}\right)\right) \qquad \text{(Using (1.6.2)}$$

$$=\frac{1}{2}\left(x+\sqrt{x^2-1}\right)^{1/n}-\frac{1}{2}\left(x+\sqrt{x^2-1}\right)^{-1/n} \qquad \text{(Using (1.6.1)}$$

$$=\frac{1}{2}\left(x+\sqrt{x^2-1}\right)^{1/n}-\frac{1}{2}\left(x-\sqrt{x^2-1}\right)^{1/n} \quad n\in R, x\geq 1.$$ (1.6.3)

Therefore, substituting n by 2^n, where n = 1, 2, 3, …, we obtain the composition of the hyperbolic sine with inverse hyperbolic cosine takes the form:

$$\sinh\left(\frac{1}{2^n}\cosh^{-1}x\right)=\frac{1}{2}\left(x+\sqrt{x^2-1}\right)^{1/2^n}-\frac{1}{2}\left(x-\sqrt{x^2-1}\right)^{1/2^n} \quad n\in N, x\geq 1.$$ (1.6.3 a)

1. Square Root Nested Radical Subtraction Identity

By combining (1.6.3a) when n = 1 and (1.5.2) we obtain a beautiful subtraction identity,

$$\sqrt{x+\sqrt{x^2-1}}-\sqrt{x-\sqrt{x^2-1}}=\sqrt{2x-2}.$$ (1.6.4)

2. Quarter Nested Radical Subtraction Identity

The derivation of the quarter nested radical subtraction identity involves the combination of (1.6.3) with n = 2 and (1.5.3), yielding the following outcome:

$$\sqrt[4]{x+\sqrt{x^2-1}}-\sqrt[4]{x-\sqrt{x^2-1}}=\sqrt{\sqrt{2+2x}-2}.$$ (1.6.5)

3. The 8th, 16th, 32th, ..., 2^{nth}, Nested Radical Subtraction Expressions

The process of deriving the nested radical subtraction expressions for the eighth, sixteenth, and subsequent powers of 2 involves combining (163) with varying values of n, ranging from 3 to n, along with (1.5.4), (1.5.5), ...(1.5.9), as demonstrated in the followings:

$$\sqrt[8]{x+\sqrt{x^2-1}}-\sqrt[8]{x-\sqrt{x^2-1}}=\sqrt{\sqrt{2+\sqrt{2+2x}}-2},$$ (1.6.6)

$$\sqrt[16]{x+\sqrt{x^2-1}}-\sqrt[16]{x-\sqrt{x^2-1}}=\sqrt{\sqrt{2+\sqrt{2+\sqrt{2+2x}}}-2},$$ (1.6.7)

$$\sqrt[32]{x+\sqrt{x^2-1}}-\sqrt[32]{x-\sqrt{x^2-1}}=\sqrt{\sqrt{2+\sqrt{2+\sqrt{2+2x}}}-2},$$ (1.6.8)

$$\sqrt[64]{x+\sqrt{x^2-1}}-\sqrt[64]{x-\sqrt{x^2-1}}=\sqrt{\sqrt{2+\sqrt{2+\sqrt{2+\sqrt{2+2x}}}}-2},$$ (1.6.9)

$$\sqrt[128]{x+\sqrt{x^2-1}}-\sqrt[128]{x-\sqrt{x^2-1}}=\sqrt{\sqrt{2+\sqrt{2+\sqrt{2+\sqrt{2+\sqrt{2+2x}}}}}-2},$$ (1.6.10)

...

$$\sqrt[2^n]{x+\sqrt{x^2-1}}-\sqrt[2^n]{x-\sqrt{x^2-1}}=\sqrt{\sqrt{2+...\sqrt{2+\sqrt{2+\sqrt{2+\sqrt{2+\sqrt{2+2x}}}}}}-2}.$$ (1.6.11)

4. Alternative Form of Nested Radical Subtraction Identities (Type 1)

The alternative expression for (1.6.4) exhibits enhancing the symmetry and aesthetic appeal of a mathematical identity when we represent x using real values a and b through the substitution $x=\sqrt{\dfrac{b}{2a}+\dfrac{1}{2}}$, leading to the subsequent formulations:

4A. Alternative Form of Nested Radical Subtraction Identities (Type 1)

a. Upon applying the x substitution, the identity of (1.6.4) becomes:

$$\sqrt{\sqrt{\frac{a}{b}+\frac{1}{2}}+\sqrt{\frac{a}{b}-\frac{1}{2}}}-\sqrt{\sqrt{\frac{a}{b}+\frac{1}{2}}-\sqrt{\frac{a}{b}-\frac{1}{2}}}=\sqrt{2\left(\sqrt{\frac{a}{b}+\frac{1}{2}}-1\right)} \qquad (1.6.12)$$

b. By further simplification, we derive:

$$\sqrt{\sqrt{a+b}+\sqrt{a-b}}-\sqrt{\sqrt{a+b}-\sqrt{a-b}}=\sqrt{2\left(\sqrt{a+b}-\sqrt{2b}\right)} \qquad (1.6.13)$$

4B. Alternative Form of Quarter Nested Radical Subtraction Identity (Type 1)

a. Similar to the concept in (1.6.5) and upon applying the x substitution, we have

$$\sqrt[4]{\sqrt{\frac{a}{b}+\frac{1}{2}}+\sqrt{\frac{a}{b}-\frac{1}{2}}}-\sqrt[4]{\sqrt{\frac{a}{b}+\frac{1}{2}}-\sqrt{\frac{a}{b}-\frac{1}{2}}}=\sqrt{\sqrt{2+2\sqrt{\frac{a}{2b}+\frac{1}{2}}}-2}$$

b. Simplifying further, we obtain

$$\sqrt[4]{\sqrt{a+b}+\sqrt{a-b}}-\sqrt[4]{\sqrt{a+b}-\sqrt{a-b}}=\sqrt{\sqrt{2\left(\sqrt{2b}+\sqrt{a+b}\right)}-2\sqrt[4]{2b}} \qquad (1.6.14)$$

4C. Alternative Form of Nested Radical Subtraction Identities for Higher Roots (Type 1)

This approach similarly yields alternative formulations for (1.6.6) through (1.6.11), we find:

$$\sqrt[8]{\sqrt{a+b}+\sqrt{a-b}}-\sqrt[8]{\sqrt{a+b}-\sqrt{a-b}}=\sqrt{\sqrt{2\sqrt[4]{2b}+\sqrt{2\left(\sqrt{2b}+\sqrt{a+b}\right)}}-2\sqrt[8]{2b}} \qquad (1.6.15)$$

$$\sqrt[16]{\sqrt{a+b}+\sqrt{a-b}}-\sqrt[16]{\sqrt{a+b}-\sqrt{a-b}}=\sqrt{\sqrt{2\sqrt[8]{2b}+\sqrt{2\sqrt[4]{2b}+\sqrt{2\left(\sqrt{2b}+\sqrt{a+b}\right)}}}-2\sqrt[16]{2b}} \qquad (1.6.16)$$

$$\sqrt[32]{\sqrt{a+b}+\sqrt{a-b}}-\sqrt[32]{\sqrt{a+b}-\sqrt{a-b}}=\sqrt{\sqrt{2\sqrt[16]{2b}+\sqrt{2\sqrt[8]{2b}+\sqrt{2\sqrt[4]{2b}+\sqrt{2\left(\sqrt{2b}+\sqrt{a+b}\right)}}}}-2\sqrt[32]{2b}} \qquad (1.6.17)$$

...

$$\sqrt[2^n]{\sqrt{a+b}+\sqrt{a-b}}-\sqrt[2^n]{\sqrt{a+b}-\sqrt{a-b}}=\sqrt{\sqrt{2\left(\sqrt[2^{n-1}]{2b}\right)+...\sqrt{2\sqrt[4]{2b}+\sqrt{2\left(\sqrt{2b}+\sqrt{a+b}\right)}}}-2\sqrt[2^n]{2b}} \qquad (1.6.18)$$

4D. Alternative Form of Nested Radical Subtraction Identities (Type 2)

Assuming that A and B are real numbers, we show the nested radical subtraction identities spanning from (1.3.4) to (1.3.14) can be reformulated as follows:

i. Define $A=\sqrt{a+b}$ and $B=\sqrt{a-b}$.

ii. Upon solving for a and b in relation to A and B, we derive:

$$a = \frac{A^2 + B^2}{2}$$

and

$$b = \frac{A^2 - B^2}{2}$$

iii. By substituting a and b into the identities (1.6.13) to (1.6.18), we obtain the following remarkably second alternative form of nested radical subtraction identities:

$$\sqrt{A+B} - \sqrt{A-B} = \sqrt{2\sqrt{A^2-B^2} - 2A} \qquad (1.6.19)$$

$$\sqrt[4]{A+B} - \sqrt[4]{A-B} = \sqrt{\sqrt{2A + 2\sqrt{A^2-B^2}} - 2\sqrt[4]{A^2-B^2}} \qquad (1.6.20)$$

$$\sqrt[8]{A+B} - \sqrt[8]{A-B} = \sqrt{\sqrt{2\sqrt[4]{A^2-B^2} + \sqrt{2A + 2\sqrt{A^2-B^2}}} - 2\sqrt[8]{A^2-B^2}} \qquad (1.6.21)$$

$$\sqrt[16]{A+B} - \sqrt[16]{A-B} = \sqrt{\sqrt{2\sqrt[8]{A^2-B^2} + \sqrt{2\sqrt[4]{A^2-B^2} + \sqrt{2A + 2\sqrt{A^2-B^2}}}} - 2\sqrt[16]{A^2-B^2}} \qquad (1.6.22)$$

...

$$\sqrt[2^n]{A+B} - \sqrt[2^n]{A-B} = \sqrt{\sqrt{2\sqrt[2^{n-1}]{A^2-B^2} + \dots \sqrt{2A + 2\sqrt{2A + 2\sqrt{A^2-B^2}}}} - 2\sqrt[2^n]{A^2-B^2}} \qquad (1.6.23)$$

Note: Formula (1.6.23) comprises n layers of nested square roots. Each successive layer builds upon the previous one by including the difference between the squares of A and B.

Section 1-7. Summary on Addition and Subtraction Identities

We've established several algebraic identities called nested radical addition in Section 1-3 and nested radical subtraction in Section 1-6. We also demonstrated the enhancement of aesthetic appeal and symmetry within identities through variable transformations. Within this section, we consolidate the identities acquired from the 1-3 and 1-6 sections. Subsequently, we illustrate the process of deriving fresh (distinct) identities by employing elementary addition or subtraction operations on existing ones.

1. Nested Radical Addition and Subtraction Identities

The nested radical addition and subtraction identities can be integrated from the 1-3 and 1-6 sections into a single comprehensive expression through the utilization of both plus and minus operations, resulting in a succinct summary:

$$\sqrt{\sqrt{a+b}+\sqrt{a-b}}\pm\sqrt{\sqrt{a+b}-\sqrt{a-b}}=\sqrt{2\left(\sqrt{a+b}\pm\sqrt{2b}\right)}, \tag{1.7.1}$$

$$\sqrt[4]{\sqrt{a+b}+\sqrt{a-b}}\pm\sqrt[4]{\sqrt{a+b}-\sqrt{a-b}}=\sqrt{\sqrt{2\left(\sqrt{2b}+\sqrt{a+b}\right)}\pm2\sqrt[4]{2b}}, \tag{1.7.2}$$

$$\sqrt[8]{\sqrt{a+b}+\sqrt{a-b}}\pm\sqrt[8]{\sqrt{a+b}-\sqrt{a-b}}=\sqrt{\sqrt{2\sqrt[4]{2b}+\sqrt{2\left(\sqrt{2b}+\sqrt{a+b}\right)}}\pm2\sqrt[8]{2b}}, \tag{1.7.3}$$

$$\sqrt[32]{\sqrt{a+b}+\sqrt{a-b}}\pm\sqrt[32]{\sqrt{a+b}-\sqrt{a-b}}=\sqrt{\sqrt{2\sqrt[16]{2b}+\sqrt{2\sqrt[8]{2b}+\sqrt{2\sqrt[4]{2b}+\sqrt{2\left(\sqrt{2b}+\sqrt{a+b}\right)}}}}\pm2\sqrt[32]{2b}}, \tag{1.7.4}$$

...

$$\sqrt[2^n]{\sqrt{a+b}+\sqrt{a-b}}\pm\sqrt[2^n]{\sqrt{a+b}-\sqrt{a-b}}=\sqrt{\sqrt{2^{2^{n-1}}\sqrt{2b}+\sqrt{2^{2^{n-2}}\sqrt{2b}+...+\sqrt{2\sqrt[4]{2b}+\sqrt{2\left(\sqrt{2b}+\sqrt{a+b}\right)}}}}\pm2\sqrt[2^n]{2b}}. \tag{1.7.5}$$

2. Summary on Alternative Form of Addition and Subtraction Identities

The alternative form of nested radical addition and subtraction are integrated from the 1-3 and 1-6 sections and summarized as follows:

$$\sqrt{A+B}\pm\sqrt{A-B}=\sqrt{2\sqrt{A^2-B^2}\pm2A}, \tag{1.7.6}$$

$$\sqrt[4]{A+B}\pm\sqrt[4]{A-B}=\sqrt{\sqrt{2A+2\sqrt{A^2-B^2}}\pm2\sqrt[4]{A^2-B^2}}, \tag{1.7.7}$$

$$\sqrt[8]{A+B}\pm\sqrt[8]{A-B}=\sqrt{\sqrt{2\sqrt[4]{A^2-B^2}+\sqrt{2A+2\sqrt{A^2-B^2}}}\pm2\sqrt[8]{A^2-B^2}}, \tag{1.7.8}$$

$$\sqrt[16]{A+B}\pm\sqrt[16]{A-B}=\sqrt{\sqrt{2\sqrt[8]{A^2-B^2}+\sqrt{2\sqrt[4]{A^2-B^2}+\sqrt{2A+2\sqrt{A^2-B^2}}}}\pm2\sqrt[16]{A^2-B^2}}, \tag{1.7.9}$$

$$\sqrt[32]{A+B}+\sqrt[32]{A-B}=\sqrt{\sqrt{2\sqrt[16]{A^2-B^2}+\sqrt{2\sqrt[8]{A^2-B^2}+\sqrt{2\sqrt[4]{A^2-B^2}+\sqrt{2A+2\sqrt{A^2-B^2}}}}}-2\sqrt[32]{A^2-B^2}}, \tag{1.7.10}$$

...

$$\sqrt[2^n]{A+B}\pm\sqrt[2^n]{A-B}=\sqrt{\sqrt{2^{2^{n-1}}\sqrt{A^2-B^2}+...\sqrt{...+\sqrt{2A+2\sqrt{A^2-B^2}}}}\pm2\sqrt[2^n]{A^2-B^2}}, \tag{1.7.11}$$

where a, b, A, and B are arbitrary.

3. Exploring Novel Nested Radical Identities Generated through Addition and Subtraction

From the earlier provided summary (1) and (2), it's apparent that each identity encompassed within the range of (1.7.1) to (1.7.11) is endowed with the ability to give rise to additional identities when combined through addition or subtraction as an illustration:

By adding two identities from (1.7.1) in part 1 of summary:

$$\sqrt{2\left(\sqrt{a+b}+\sqrt{2b}\right)}+\sqrt{2\left(\sqrt{a+b}-\sqrt{2b}\right)}=2\sqrt{\sqrt{a+b}+\sqrt{a-b}}$$

By subtracting them from each other:

$$\sqrt{2\left(\sqrt{a+b}+\sqrt{2b}\right)}-\sqrt{2\left(\sqrt{a+b}-\sqrt{2b}\right)}=2\sqrt{\sqrt{a+b}-\sqrt{a-b}}$$

Upon consolidation, a novel addition and subtraction identity emerges, resembling the original ones, namely

$$\sqrt{2\left(\sqrt{a+b}+\sqrt{2b}\right)}\pm\sqrt{2\left(\sqrt{a+b}-\sqrt{2b}\right)}=2\sqrt{\sqrt{a+b}\pm\sqrt{a-b}} \qquad (1.7.12)$$

Similar to the above concept, which can be applied to expressions (1.7.2), (1.7.8), and (1.7.9) in part 2 of the summary, we find:

$$\sqrt{\sqrt{2\left(\sqrt{2b}+\sqrt{a+b}\right)}+2\sqrt[4]{2b}}\pm\sqrt{\sqrt{2\left(\sqrt{2b}+\sqrt{a+b}\right)}-2\sqrt[4]{2b}}=\sqrt[4]{\sqrt{a+b}\pm\sqrt{a-b}}. \qquad (1.7.13)$$

$$\sqrt{\sqrt{2\sqrt[4]{A^2-B^2}+\sqrt{2A+2\sqrt{A^2-B^2}}}+2\sqrt[8]{A^2-B^2}}\pm\sqrt{\sqrt{2\sqrt[4]{A^2-B^2}+\sqrt{2A+2\sqrt{A^2-B^2}}}-2\sqrt[8]{A^2-B^2}}=\sqrt[8]{A\pm B} \qquad (1.7.14)$$

and

$$\sqrt{\sqrt{2\sqrt[8]{A^2-B^2}+\sqrt{2\sqrt[4]{A^2-B^2}+\sqrt{2A+2\sqrt{A^2-B^2}}}}+2\sqrt[16]{A^2-B^2}}$$
$$\pm\sqrt{\sqrt{2\sqrt[8]{A^2-B^2}+\sqrt{2\sqrt[4]{A^2-B^2}+\sqrt{2A+2\sqrt{A^2-B^2}}}}-2\sqrt[16]{A^2-B^2}}=\sqrt[16]{A\pm B}. \qquad (1.7.15)$$

Evidently, we can formulate several distinct nested radical addition and subtraction identities within the realm of even radicals. Notably, cubic or quintic radicals do not manifest in this context, as it involves solely a complex radical that incorporates the complex unit i. However, Srinivasa Ramanujan did manage to generate certain cubic nested radical expressions, some of which we shall introduce in the next section.

Section 1-8. Nested Radical Generated by Composition cos[(1/2ⁿ)sin⁻¹(x)]

In this section, we will establish the nested radical identities associated with the cosine and inverse sine with the structure:

$$\cos\left(\frac{1}{2^n}\sin^{-1}(x)\right)\text{ with }-1\leq x\leq 1,\text{ and }n\in N=\{1,2,3,\ldots\}.$$

We have $\sin\left(\cos^{-1}(x)\right)=\sqrt{1-x^2}$. [see Appendix 1-A.]

Equivalently, it can be expressed as

$$\cos^{-1}(x)=\sin^{-1}\left(\sqrt{1-x^2}\right)\text{ for }0\leq x\leq 1. \tag{1.8.1}$$

and

$$\cos^{-1}(-x)=\sin^{-1}\left(\sqrt{1-x^2}\right)\text{ for }-1\leq x\leq 0. \tag{1.8.1*}$$

By substituting expression $\cos^{-1}(x)$ from (1.8.1) into (1.2), we have

$$\cos\left(\frac{1}{2}\sin^{-1}\left(\sqrt{1-x^2}\right)\right)=\frac{1}{2}\sqrt{2+2x} \tag{1.8.2}$$

Setting

$$u=\sqrt{1-x^2}$$

$$\Rightarrow x=+\sqrt{1-u^2}\text{ for }-1\leq u\leq 1.$$

Then substituting x into (1.9.1), we obtain

$$\cos\left(\frac{1}{2}\sin^{-1}(u)\right)=\frac{1}{2}\sqrt{2+2\sqrt{1-u^2}}.$$

Rather than employing the variable u in the identity, by substituting u with x, we achieve the intended outcome:

$$\cos\left(\frac{1}{2}\sin^{-1}(x)\right)=\frac{1}{2}\sqrt{2+2\sqrt{1-x^2}},\ -1\leq x\leq 1 \tag{1.8.3}$$

Applying the double-angle cosine formula to the left-hand side of (1.8.3) yields

$$2\cos^2\left(\frac{1}{4}\sin^{-1}(x)\right)-1=\frac{1}{2}\sqrt{2+2\sqrt{1-x^2}}$$

Hence, by simplifying further, we deduce that

$$\cos\left(\frac{1}{4}\sin^{-1}x\right)=\frac{1}{2}\sqrt{2+\sqrt{2+2\sqrt{1-x^2}}}, \quad -1\le x\le 1 \tag{1.8.4}$$

Note that the negative square root of (1.8.4) is disregarded, as the right-hand side remains positive for -1 ≤ x ≤ 1.

Continuing the process in the manner previously described, we derive subsequent identities as presented below:

$$\cos\left(\frac{1}{8}\sin^{-1}(x)\right)=\frac{1}{2}\sqrt{2+\sqrt{2+\sqrt{2+2\sqrt{1-x^2}}}}, \quad -1\le x\le 1 \tag{1.8.5}$$

$$\cos\left(\frac{1}{16}\sin^{-1}(x)\right)=\frac{1}{2}\sqrt{2+\sqrt{2+\sqrt{2+\sqrt{2+2\sqrt{1-x^2}}}}}, \quad -1\le x\le 1 \tag{1.8.6}$$

$$\cos\left(\frac{1}{32}\sin^{-1}(x)\right)=\frac{1}{2}\sqrt{2+\sqrt{2+\sqrt{2+\sqrt{2+\sqrt{2+2\sqrt{1-x^2}}}}}}, \quad -1\le x\le 1 \tag{1.8.7}$$

$$\cos\left(\frac{1}{64}\sin^{-1}(x)\right)=\frac{1}{2}\sqrt{2+\sqrt{2+\sqrt{2+\sqrt{2+\sqrt{2+\sqrt{2+2\sqrt{1-x^2}}}}}}}, \quad -1\le x\le 1 \tag{1.8.8}$$

$$\cos\left(\frac{1}{128}\sin^{-1}(x)\right)=\frac{1}{2}\sqrt{2+\sqrt{2+\sqrt{2+\sqrt{2+\sqrt{2+\sqrt{2+\sqrt{2+2\sqrt{1-x^2}}}}}}}}, \quad -1\le x\le 1 \tag{1.8.9}$$

...

$$\cos\left(\frac{1}{2^n}\sin^{-1}(x)\right)=\frac{1}{2}\sqrt{2+...\sqrt{2+\sqrt{2+\sqrt{2+\sqrt{2+\sqrt{2+\sqrt{2+2\sqrt{1-x^2}}}}}}}}, \quad -1\le x\le 1 \tag{1.8.10}$$

Section 1-9. Nested Radical Generated by Composition cosh[(1/2ⁿ)sinh⁻¹(x)]

In this section, we will establish the nested radical identities associated with the hyperbolic cosine and inverse hyperbolic sine in the form:

$$\cosh\left(\frac{1}{2^n}\sinh^{-1}(x)\right) \text{ for } -\infty < x < \infty \text{ and } n \in \mathbb{N} = \{1,2,3,\dots\} \tag{1.9.1}$$

where the $\sinh^{-1}(x)$ is the inverse function of $\sinh(x)$, and is defined as

$$\sinh^{-1}(x) = \ln\left(x + \sqrt{x^2+1}\right), -\infty \le x < \infty.$$

The formula (1.9.1) can be derived from the definition as follows:

$$\cosh\left(\frac{1}{2^n}\sinh^{-1}(x)\right) = \cosh\left(\frac{1}{2^n}\ln\left(x+\sqrt{x^2+1}\right)\right)$$

$$= \frac{1}{2}\left[\left(x+\sqrt{x^2+1}\right)^{1/2^n} + \left(x+\sqrt{x^2+1}\right)^{-1/2^n}\right] \tag{1.9.1a}$$

$$= \frac{1}{2}\left[\left(\sqrt{x^2+1}+x\right)^{1/2^n} + \left(\sqrt{x^2+1}-x\right)^{-1/2^n}\right] \tag{1.9.1b}$$

On the other hand, for real x and by substituting x with ix in each identity ranging from (1.8.3) to (1.8.10) then applying $\sin^{-1}(ix) = i\sinh^{-1}(x)$ and $\cos(ix) = \cosh(x)$, we derive the subsequent identities outlined below:

$$\cosh\left(\frac{1}{2}\sinh^{-1}(x)\right) = \frac{1}{2}\sqrt{2+2\sqrt{1+x^2}}, \quad -\infty < x < \infty \tag{1.9.2}$$

$$\cosh\left(\frac{1}{4}\sinh^{-1}(x)\right) = \frac{1}{2}\sqrt{2+\sqrt{2+2\sqrt{1+x^2}}}, \quad -\infty < x < \infty \tag{1.9.3}$$

$$\cosh\left(\frac{1}{8}\sinh^{-1}(x)\right) = \frac{1}{2}\sqrt{2+\sqrt{2+\sqrt{2+2\sqrt{1+x^2}}}}, \quad -\infty < x < \infty \tag{1.9.4}$$

$$\cosh\left(\frac{1}{16}\sinh^{-1}(x)\right) = \frac{1}{2}\sqrt{2+\sqrt{2+\sqrt{2+\sqrt{2+2\sqrt{1+x^2}}}}}, \quad -\infty < x < \infty \tag{1.9.5}$$

$$\cosh\left(\frac{1}{32}\sinh^{-1}(x)\right) = \frac{1}{2}\sqrt{2+\sqrt{2+\sqrt{2+\sqrt{2+\sqrt{2+2\sqrt{1+x^2}}}}}}, \quad -\infty < x < \infty \tag{1.9.6}$$

$$\cosh\left(\frac{1}{64}\sinh^{-1}(x)\right) = \frac{1}{2}\sqrt{2+\sqrt{2+\sqrt{2+\sqrt{2+\sqrt{2+\sqrt{2+2\sqrt{1+x^2}}}}}}}, \quad -\infty < x < \infty \tag{1.9.7}$$

$$\cosh\left(\frac{1}{128}\sinh^{-1}(x)\right)=\frac{1}{2}\sqrt{2+\sqrt{2+\sqrt{2+\sqrt{2+\sqrt{2+\sqrt{2+\sqrt{2+2\sqrt{1+x^2}}}}}}}},\ -\infty<x<\infty \qquad (1.9.8)$$

...

$$\cosh\left(\frac{1}{2^n}\sinh^{-1}(x)\right)=\frac{1}{2}\sqrt{2+...+\sqrt{2+\sqrt{2+\sqrt{2+\sqrt{2+\sqrt{2+2\sqrt{1+x^2}}}}}}}.\ -\infty<x<\infty \qquad (1.9.9)$$

By employing the outcomes derived from (1.9.2) through (1.9.9) in conjunction with formula (1.9.1b) for n = 1, 2, 3,..., we obtain the results presented below:

$$\sqrt{\sqrt{x^2+1}+x}+\sqrt{\sqrt{x^2+1}-x}=\sqrt{2+2\sqrt{1+x^2}},\ -\infty<x<\infty \qquad (1.9.10)$$

$$\sqrt[4]{\sqrt{x^2+1}+x}+\sqrt[4]{\sqrt{x^2+1}-x}=\sqrt{2+\sqrt{2+2\sqrt{1+x^2}}},\ -\infty<x<\infty \qquad (1.9.11)$$

$$\sqrt[8]{\sqrt{x^2+1}+x}+\sqrt[8]{\sqrt{x^2+1}-x}=\sqrt{2+\sqrt{2+\sqrt{2+2\sqrt{1+x^2}}}},\ -\infty<x<\infty \qquad (1.9.12)$$

...

$$\sqrt[2^n]{\sqrt{x^2+1}+x}+\sqrt[2^n]{\sqrt{x^2+1}-x}=\sqrt{2+...\sqrt{2+\sqrt{2+\sqrt{2+\sqrt{2+\sqrt{2+2\sqrt{1+x^2}}}}}}},\ -\infty<x<\infty \qquad (1.9.13)$$

Section 1-10. Nested Radical Generated by Composition cos[(1/2ⁿ)tan⁻¹(x)]

In this section, we will establish the nested radical identities associated with the cosine and inverse tangent in the form:

$$\cos\left(\frac{1}{2^n}\tan^{-1}(x)\right),\ x\in\mathbb{R},\ n\in\mathbb{N}=\{1,2,3,...\}. \qquad (1.10.1)$$

We know that

$$\tan\left(\sin^{-1}(x)\right)=\frac{x}{\sqrt{1-x^2}}.$$

or

$$\sin^{-1}(x) = \tan^{-1}\left(\frac{x}{\sqrt{1-x^2}}\right)$$

Substituting the expression for $\sin^{-1}(x)$ into (1.2), we have

$$\cos\left(\frac{1}{2}\tan^{-1}\left(\frac{x}{\sqrt{1-x^2}}\right)\right) = \frac{1}{2}\sqrt{2+2\sqrt{1-x^2}} \tag{1.10.2}$$

Setting

$$\frac{x}{\sqrt{1-x^2}} = u$$

$$\Rightarrow x^2 = u^2 - u^2 x^2$$

$$\Rightarrow x^2 = \frac{u^2}{1+u^2}.$$

Substituting the expression for x^2 into (1.10.2) yields

$$\cos\left(\frac{1}{2}\tan^{-1}(u)\right) = \sqrt{\frac{1}{2} + \frac{1}{2\sqrt{1+u^2}}} = \frac{1}{2}\sqrt{2 + \frac{2}{\sqrt{1+u^2}}}, \quad -\infty < u < \infty.$$

By renaming u as x, we obtain

$$\cos\left(\frac{1}{2}\tan^{-1}(x)\right) = \frac{1}{2}\sqrt{2 + \frac{2}{\sqrt{1+x^2}}}, \quad -\infty < x < \infty. \tag{1.10.3}$$

Applying the double-angle cosine formula to the left-hand side of (1.10.3) gives

$$2\cos^2\left(\frac{1}{2}\tan^{-1}(x)\right) - 1 = \frac{1}{2}\sqrt{2 + \frac{2}{\sqrt{1+x^2}}}$$

Consequently, by simplifying further, we obtain

$$\cos\left(\frac{1}{4}\tan^{-1}(x)\right) = \frac{1}{2}\sqrt{2 + \sqrt{2 + \frac{2}{\sqrt{1+x^2}}}}, \quad -\infty < x < \infty. \tag{1.10.4}$$

Continuing the process in the manner previously described, we derive subsequent identities as presented below:

$$\cos\left(\frac{1}{8}\tan^{-1}(x)\right)=\frac{1}{2}\sqrt{2+\sqrt{2+\sqrt{2+\frac{2}{\sqrt{1+x^2}}}}}, \quad -\infty<x<\infty \tag{1.10.5}$$

$$\cos\left(\frac{1}{16}\tan^{-1}(x)\right)=\frac{1}{2}\sqrt{2+\sqrt{2+\sqrt{2+\sqrt{2+\frac{2}{\sqrt{1+x^2}}}}}}, \quad -\infty<x<\infty \tag{1.10.6}$$

$$\cos\left(\frac{1}{32}\tan^{-1}(x)\right)=\frac{1}{2}\sqrt{2+\sqrt{2+\sqrt{2+\sqrt{2+\sqrt{2+\frac{2}{\sqrt{1+x^2}}}}}}}, \quad -\infty<x<\infty \tag{1.10.7}$$

$$\cos\left(\frac{1}{64}\tan^{-1}(x)\right)=\frac{1}{2}\sqrt{2+\sqrt{2+\sqrt{2+\sqrt{2+\sqrt{2+\sqrt{2+\frac{2}{\sqrt{1+x^2}}}}}}}}, \quad -\infty<x<\infty \tag{1.10.8}$$

$$\cos\left(\frac{1}{128}\tan^{-1}(x)\right)=\frac{1}{2}\sqrt{2+\sqrt{2+\sqrt{2+\sqrt{2+\sqrt{2+\sqrt{2+\sqrt{2+\frac{2}{\sqrt{1+x^2}}}}}}}}}, \quad -\infty<x<\infty \tag{1.10.9}$$

...

$$\cos\left(\frac{1}{2^n}\tan^{-1}(x)\right)=\frac{1}{2}\sqrt{2+...\sqrt{2+\sqrt{2+\sqrt{2+\sqrt{2+\sqrt{2+\sqrt{2+\frac{2}{\sqrt{1+x^2}}}}}}}}}, \quad -\infty<x<\infty \tag{1.10.10}$$

Section 1-11. Nested Radical Generated by Composition cosh[(1/2ⁿ)tanh⁻¹(x)]

We establish the nested radical identities associated with the hyperbolic cosine and inverse tangent with the structure:

$$\cosh\left(\frac{1}{2^n}\tanh^{-1}(x)\right), \quad -1\leq x\leq 1, \quad n\in I\!\!N=\{1,2,3,...\}, \tag{1.11.1}$$

where

$$\tanh^{-1}(x) = \frac{1}{2} \ln \frac{1+x}{1-x}, -1 < x < 1.$$

The formula (1.11.1) can be explicitly rewritten as

$$\cosh\left(\frac{1}{2^n} \tanh^{-1}(x)\right) = \cosh\left(\frac{1}{2^n} \cdot \frac{1}{2} \ln \frac{1+x}{1-x}\right), -1 < x < 1$$

$$= \frac{1}{2}\left[\left(\frac{1+x}{1-x}\right)^{\frac{1}{2^{n+1}}} + \left(\frac{1+x}{1-x}\right)^{-\frac{1}{2^{n+1}}}\right], \quad -1 < x < 1 \tag{1.11.1a}$$

$$= \frac{1}{2}\left[\left(\frac{1+x}{1-x}\right)^{\frac{1}{2^{n+1}}} + \left(\frac{1-x}{1+x}\right)^{\frac{1}{2^{n+1}}}\right], \quad -1 < x < 1 \tag{1.11.1b}$$

On the other hand, upon substituting x with ix in each identity ranging from (1.10.3) to (1.10.10), then applying $\tan^{-1}(ix) = i\tanh^{-1}(x)$ and $\cos(ix) = \cosh(x)$, we derive the subsequent identities outlined below:

$$\cosh\left(\frac{1}{2} \tanh^{-1}(x)\right) = \frac{1}{2}\sqrt{2 + \frac{2}{\sqrt{1-x^2}}}, \quad -1 < x < 1 \tag{1.11.2}$$

$$\cosh\left(\frac{1}{4} \tanh^{-1}(x)\right) = \frac{1}{2}\sqrt{2 + \sqrt{2 + \frac{2}{\sqrt{1-x^2}}}}, \quad -1 < x < 1 \tag{1.11.3}$$

$$\cosh\left(\frac{1}{16} \tanh^{-1}(x)\right) = \frac{1}{2}\sqrt{2 + \sqrt{2 + \sqrt{2 + \sqrt{2 + \frac{2}{\sqrt{1-x^2}}}}}}, \quad -1 < x < 1 \tag{1.11.4}$$

$$\cosh\left(\frac{1}{32} \tanh^{-1}(x)\right) = \frac{1}{2}\sqrt{2 + \sqrt{2 + \sqrt{2 + \sqrt{2 + \sqrt{2 + \frac{2}{\sqrt{1-x^2}}}}}}}, \quad -1 < x < 1 \tag{1.11.5}$$

$$\cosh\left(\frac{1}{64} \tanh^{-1}(x)\right) = \frac{1}{2}\sqrt{2 + \sqrt{2 + \sqrt{2 + \sqrt{2 + \sqrt{2 + \sqrt{2 + \frac{2}{\sqrt{1-x^2}}}}}}}}, \quad -1 < x < 1 \tag{1.11.6}$$

$$\cosh\left(\frac{1}{128}\tanh^{-1}(x)\right)=\frac{1}{2}\sqrt{2+\sqrt{2+\sqrt{2+\sqrt{2+\sqrt{2+\sqrt{2+\sqrt{2+\frac{2}{\sqrt{1-x^2}}}}}}}}},\ -1<x<1 \tag{1.11.7}$$

...

$$\cosh\left(\frac{1}{2^n}\tanh^{-1}(x)\right)=\frac{1}{2}\sqrt{2+...\sqrt{2+\sqrt{2+\sqrt{2+\sqrt{2+\sqrt{2+\sqrt{2+\frac{2}{\sqrt{1-x^2}}}}}}}}},\ -1<x<1 \tag{1.11.8}$$

By employing the outcomes derived from (1.11.2) through (1.11.8) in conjunction with formula (1.11.1b) for n = 1, 2, 3,..., we obtain the addition identities:

$$\sqrt[4]{\frac{1+x}{1-x}}+\sqrt[4]{\frac{1-x}{1+x}}=\sqrt{2+\frac{2}{\sqrt{1-x^2}}},-1<x<1 \tag{1.11.9}$$

$$\sqrt[8]{\frac{1+x}{1-x}}+\sqrt[8]{\frac{1-x}{1+x}}=\sqrt{2+\sqrt{2+\frac{2}{\sqrt{1-x^2}}}},\ -1<x<1 \tag{1.11.10}$$

$$\sqrt[16]{\frac{1+x}{1-x}}+\sqrt[16]{\frac{1-x}{1+x}}=\sqrt{2+\sqrt{2+\sqrt{2+\sqrt{2+\frac{2}{\sqrt{1-x^2}}}}}},\ -1<x<1 \tag{1.11.11}$$

...

$$\sqrt[2^{n+1}]{\frac{1+x}{1-x}}+\sqrt[2^{n+1}]{\frac{1-x}{1+x}}=\sqrt{2+...+\sqrt{2+\sqrt{2+\sqrt{2+\sqrt{2+\sqrt{2+\sqrt{2+\frac{2}{\sqrt{1-x^2}}}}}}}}},\ -1<x<1 \tag{1.11.12}$$

Section 1-12. Nested Radical Generated by Composition sin[(1/2ⁿ)tan⁻¹(x)], and sinh[(1/2ⁿ)tanh⁻¹(x)]

1. Composition of Sine with Inverse Tangent - sin[(1/2ⁿ)tan⁻¹(x)]

The nested radical expressions can be formed by composing sine function and inverse tangent function with the structure:

$$\sin\left(\frac{1}{2^n}\tan^{-1}(x)\right),\quad x\in\mathbb{R},\quad n\in\mathbb{N}=\{1,2,3,\ldots\}. \tag{1.12.1}$$

Squaring both sides of (1.10.3), we have

$$\cos^2\left(\frac{1}{2}\tan^{-1}(x)\right)=\frac{1}{4}\left(2+\frac{2}{\sqrt{1+x^2}}\right)$$

$$=\frac{1}{2}\left(1+\frac{1}{\sqrt{1+x^2}}\right)$$

$$\Leftrightarrow 1-\sin^2\left(\frac{1}{2}\tan^{-1}(x)\right)=\frac{1}{2}\left(1+\frac{1}{\sqrt{1+x^2}}\right)$$

$$\Leftrightarrow \sin^2\left(\frac{1}{2}\tan^{-1}(x)\right)=\frac{1}{2}\left(1-\frac{1}{\sqrt{1+x^2}}\right)$$

$$\Rightarrow \sin\left(\frac{1}{2}\tan^{-1}(x)\right)=\begin{cases}\dfrac{1}{2}\sqrt{2-\dfrac{2}{\sqrt{1+x^2}}} & x\geq 0 \\[4mm] -\dfrac{1}{2}\sqrt{2-\dfrac{2}{\sqrt{1+x^2}}} & x\leq 0\end{cases} \tag{1.12.2}$$

Note:

- The expression within the square root on the right-hand side of equation (1.12.2) is always greater than or equal to 0 for real values of x. Indeed, it meets the following condition:

$$\left|2-\frac{2}{\sqrt{1+x^2}}\right|\geq 0$$

$$\Leftrightarrow \left|\sqrt{1+x^2}-1\right|\geq 0$$

$$\Leftrightarrow |x|\geq 0.$$

- Although formula (1.12.2) could be rephrased using the sgn(x) function for simplicity, we have opted not to do so in this book.

To find sin((1/4) tan⁻¹(x)), we square both sides of (1.10.4), we have

$$\cos^2\left(\frac{1}{4}\tan^{-1}(x)\right)=\frac{1}{4}\left(2+\sqrt{2+\frac{2}{\sqrt{1+x^2}}}\right)$$

$$\Leftrightarrow 1 - \sin^2\left(\frac{1}{4}\tan^{-1}(x)\right) = \frac{1}{4}\left(2 + \sqrt{2 + \frac{2}{\sqrt{1+x^2}}}\right)$$

By simplifying further, we deduce that

$$\sin\left(\frac{1}{4}\tan^{-1}(x)\right) = \begin{cases} \dfrac{1}{2}\sqrt{2 - \sqrt{2 + \dfrac{2}{\sqrt{1+x^2}}}} & x \geq 0 \\[4mm] -\dfrac{1}{2}\sqrt{2 - \sqrt{2 + \dfrac{2}{\sqrt{1+x^2}}}} & x \leq 0 \end{cases} \tag{1.12.3}$$

Following the process in the manner previously described, continuing with (1.10.5) and proceed to (1.10.10), we derive the subsequent identities as shown below, respectively:

$$\sin\left(\frac{1}{8}\tan^{-1}(x)\right) = \pm\frac{1}{2}\sqrt{2 - \sqrt{2 + \sqrt{2 + \frac{2}{\sqrt{1+x^2}}}}} \tag{1.12.4}$$

$$\sin\left(\frac{1}{16}\tan^{-1}(x)\right) = \pm\frac{1}{2}\sqrt{2 - \sqrt{2 + \sqrt{2 + \sqrt{2 + \frac{2}{\sqrt{1+x^2}}}}}} \tag{1.12.5}$$

$$\sin\left(\frac{1}{32}\tan^{-1}(x)\right) = \pm\frac{1}{2}\sqrt{2 - \sqrt{2 + \sqrt{2 + \sqrt{2 + \sqrt{2 + \frac{2}{\sqrt{1+x^2}}}}}}} \tag{1.12.6}$$

$$\sin\left(\frac{1}{64}\tan^{-1}(x)\right) = \pm\frac{1}{2}\sqrt{2 - \sqrt{2 + \sqrt{2 + \sqrt{2 + \sqrt{2 + \sqrt{2 + \frac{2}{\sqrt{1+x^2}}}}}}}} \tag{1.12.7}$$

$$\sin\left(\frac{1}{128}\tan^{-1}(x)\right) = \pm\frac{1}{2}\sqrt{2 - \sqrt{2 + \sqrt{2 + \sqrt{2 + \sqrt{2 + \sqrt{2 + \sqrt{2 + \frac{2}{\sqrt{1+x^2}}}}}}}}} \tag{1.12.8}$$

30

...

$$\sin\left(\frac{1}{2^n}\tan^{-1}(x)\right)=\pm\frac{1}{2}\sqrt{2-\sqrt{2+...+\sqrt{2+\sqrt{2+\sqrt{2+\sqrt{2+\sqrt{2+\frac{2}{\sqrt{1+x^2}}}}}}}}} \qquad (1.12.9)$$

The positive sign is used for all identities when x≥0, and the negative sign is used when x<0.

2. Nested Radical Generated by Composition sinh[(1/2ⁿ)tanh⁻¹(x)]

We can derive the nested radical expressions that result from composing the hyperbolic sine function and inverse hyperbolic tangent function with the structure

$$\sinh\left(\frac{1}{2^n}\tanh^{-1}(x)\right),\ \ |x|<1,\ \ n\in N=\{1,2,3,...\}, \qquad (1.12.10)$$

where

$$\tanh^{-1}(x)=\frac{1}{2}\ln\frac{1+x}{1-x},\ -1<x<1.$$

The formula (1.12.10) can be explicitly rewritten as

$$\sinh\left(\frac{1}{2^n}\tanh^{-1}(x)\right)=\sinh\left(\frac{1}{2^n}\cdot\frac{1}{2}\ln\frac{1+x}{1-x}\right) \qquad -1<x<1,$$

$$=\frac{1}{2}\left[\left(\frac{1+x}{1-x}\right)^{\frac{1}{2^{n+1}}}-\left(\frac{1+x}{1-x}\right)^{-\frac{1}{2^{n+1}}}\right] \quad -1<x<1, \qquad (1.12.1a)$$

$$=\frac{1}{2}\left[\left(\frac{1+x}{1-x}\right)^{\frac{1}{2^{n+1}}}-\left(\frac{1-x}{1+x}\right)^{\frac{1}{2^{n+1}}}\right] \quad -1<x<1. \qquad (1.12.1b)$$

Alternatively, by applying the identities sin(ix) = isinh(x) and tan⁻¹(ix) = itanh⁻¹(x) in Part 1, we derive the results as shown below:

$$\sinh\left(\frac{1}{4}\tanh^{-1}(x)\right)=\pm\frac{1}{2}\sqrt{\frac{2}{\sqrt{1-x^2}}-2}, \qquad (1.12.2)$$

$$\sinh\left(\frac{1}{8}\tanh^{-1}(x)\right)=\pm\frac{1}{2}\sqrt{2+\frac{2}{\sqrt{1-x^2}}-2}, \qquad (1.12.3)$$

31

$$\sinh\left(\frac{1}{16}\tanh^{-1}(x)\right)=\pm\frac{1}{2}\sqrt{\sqrt{2+\sqrt{2+\frac{2}{\sqrt{1-x^2}}}}-2},$$

(1.12.4)

$$\sinh\left(\frac{1}{32}\tanh^{-1}(x)\right)=\pm\frac{1}{2}\sqrt{\sqrt{\sqrt{2+\sqrt{2+\sqrt{2+\frac{2}{\sqrt{1-x^2}}}}}}-2},$$

(1.12.5)

$$\sinh\left(\frac{1}{64}\tanh^{-1}(x)\right)=\pm\frac{1}{2}\sqrt{\sqrt{2+\sqrt{2+\sqrt{2+\sqrt{2+\frac{2}{\sqrt{1-x^2}}}}}}-2},$$

(1.12.6)

$$\sinh\left(\frac{1}{128}\tanh^{-1}(x)\right)=\pm\frac{1}{2}\sqrt{\sqrt{2+\sqrt{2+\sqrt{2+\sqrt{2+\sqrt{2+\frac{2}{\sqrt{1-x^2}}}}}}}-2}$$

(1.12.7)

...

$$\sinh\left(\frac{1}{2^n}\tanh^{-1}(x)\right)=\pm\frac{1}{2}\sqrt{\sqrt{2+...+\sqrt{2+\sqrt{2+\sqrt{2+\sqrt{2+\sqrt{2+\frac{2}{\sqrt{1-x^2}}}}}}}}-2}.$$

(1.12.8)

The positive sign is used when $0\leq x<1$, and the negative sign is used when $-1<x\leq0$.

By employing the outcomes derived from (1.12.2) through (1.12.8) in conjunction with formula (1.12.1b) for n = 1, 2, 3,..., we obtain the subtraction identities:

$$\sqrt[4]{\frac{1+x}{1-x}}-\sqrt[4]{\frac{1-x}{1+x}}=\pm\sqrt{\frac{2}{\sqrt{1-x^2}}-2},\ -1<x<1$$

(1.12.9)

$$\sqrt[8]{\frac{1+x}{1-x}}-\sqrt[8]{\frac{1-x}{1+x}}=\pm\sqrt{2+\frac{2}{\sqrt{1-x^2}}-2},\ -1<x<1$$

(1.12.10)

$$\sqrt[16]{\frac{1+x}{1-x}}-\sqrt[16]{\frac{1-x}{1+x}}=\pm\sqrt{\sqrt{2+\sqrt{2+\frac{2}{\sqrt{1-x^2}}}}-2},\ -1<x<1$$

(1.12.11)

...

$$\sqrt[2^{n+1}]{\frac{1+x}{1-x}} - \sqrt[2^{n+1}]{\frac{1-x}{1+x}} = \pm\sqrt{2+...+\sqrt{2+\sqrt{2+\sqrt{2+\sqrt{2+\sqrt{2+\frac{2}{\sqrt{1-x^2}}}}}}}} - 2, \quad -1<x<1 \tag{1.12.12}$$

(The positive sign is used for 0≤x<1, while the negative sign is used for -1<x<0.)

Section 1-13. Nested Radical Generated by Composition tan[(1/2ⁿ)tan⁻¹(x)], and Composition tanh[(1/2ⁿ)tanh⁻¹(x)]

1. Nested Radical Generated by Composition tan[(1/2ⁿ)tan⁻¹(x)]

We establish the nested radical expressions associated with the tangent function and inverse tangent function with the structure:

$$\tan\left(\frac{1}{2^n}\tan^{-1}(x)\right), \quad x\in R, \quad n\in N=\{1,2,3,...\}. \tag{1.13.1}$$

Given that the cosine and sine of $(1/2^n)\tan^{-1}(x)$ have been determined in Section 1-11 and Section 1-12, establishing the tangent function becomes straightforward. Below, we present a summary of the derived identities.

$$\tan\left(\frac{1}{2}\tan^{-1}(x)\right) = \frac{\sin\left(\frac{1}{2}\tan^{-1}(x)\right)}{\cos\left(\frac{1}{2}\tan^{-1}(x)\right)}$$

$$= \begin{cases} \sqrt{\dfrac{\sqrt{1+x^2}-1}{\sqrt{1+x^2}+1}} & x\geq 0 \\[3mm] -\sqrt{\dfrac{\sqrt{1+x^2}-1}{\sqrt{1+x^2}+1}} & x\leq 0 \end{cases} \tag{1.13.2}$$

$$\tag{1.13.3}$$

$$\tan\left(\frac{1}{4}\tan^{-1}(x)\right) = \sqrt{\frac{2\sqrt[4]{1+x^2}-\sqrt{2+2\sqrt{1+x^2}}}{2\sqrt[4]{1+x^2}+\sqrt{2+2\sqrt{1+x^2}}}}, \quad x\geq 0 \tag{1.13.3a}$$

$$\tan\left(\frac{1}{4}\tan^{-1}(x)\right)=-\sqrt{\frac{2\sqrt[4]{1+x^2}-\sqrt{2+2\sqrt{1+x^2}}}{2\sqrt[4]{1+x^2}+\sqrt{2+2\sqrt{1+x^2}}}},\quad x\le 0 \tag{1.13.3 b}$$

$$(1.13.4)$$

$$\tan\left(\frac{1}{8}\tan^{-1}(x)\right)=\sqrt{\frac{2\sqrt[8]{1+x^2}-\sqrt{\sqrt{2+2\sqrt{1+x^2}}+2\sqrt[4]{1+x^2}}}{2\sqrt[8]{1+x^2}+\sqrt{\sqrt{2+2\sqrt{1+x^2}}+2\sqrt[4]{1+x^2}}}},\quad x\ge 0. \tag{1.13.4 a}$$

$$\tan\left(\frac{1}{8}\tan^{-1}(x)\right)=-\sqrt{\frac{2\sqrt[8]{1+x^2}-\sqrt{\sqrt{2+2\sqrt{1+x^2}}+2\sqrt[4]{1+x^2}}}{2\sqrt[8]{1+x^2}+\sqrt{\sqrt{2+2\sqrt{1+x^2}}+2\sqrt[4]{1+x^2}}}},\quad x\le 0. \tag{1.13.4 b}$$

$$(1.13.5)$$

$$\tan\left(\frac{1}{16}\tan^{-1}(x)\right)=\sqrt{\frac{2\sqrt[16]{1+x^2}-\sqrt{\sqrt{\sqrt{2+2\sqrt{1+x^2}}+2\sqrt[4]{1+x^2}}+2\sqrt[8]{1+x^2}}}{2\sqrt[16]{1+x^2}+\sqrt{\sqrt{\sqrt{2+2\sqrt{1+x^2}}+2\sqrt[4]{1+x^2}}+2\sqrt[8]{1+x^2}}}},\quad x\ge 0 \tag{1.13.5 a}$$

$$\tan\left(\frac{1}{16}\tan^{-1}(x)\right)=-\sqrt{\frac{2\sqrt[16]{1+x^2}-\sqrt{\sqrt{\sqrt{2+2\sqrt{1+x^2}}+2\sqrt[4]{1+x^2}}+2\sqrt[8]{1+x^2}}}{2\sqrt[16]{1+x^2}+\sqrt{\sqrt{\sqrt{2+2\sqrt{1+x^2}}+2\sqrt[4]{1+x^2}}+2\sqrt[8]{1+x^2}}}},\quad x\le 0 \tag{1.13.5 b}$$

$$(1.13.6)$$

$$\tan\left(\frac{1}{32}\tan^{-1}(x)\right)=\sqrt{\frac{2\sqrt[32]{1+x^2}-\sqrt{\sqrt{\sqrt{\sqrt{2+2\sqrt{1+x^2}}+2\sqrt[4]{1+x^2}}+2\sqrt[8]{1+x^2}}+2\sqrt[16]{1+x^2}}}{2\sqrt[32]{1+x^2}+\sqrt{\sqrt{\sqrt{\sqrt{2+2\sqrt{1+x^2}}+2\sqrt[4]{1+x^2}}+2\sqrt[8]{1+x^2}}+2\sqrt[16]{1+x^2}}}},\quad x\ge 0 \tag{1.13.6 a}$$

$$\tan\left(\frac{1}{32}\tan^{-1}(x)\right)=-\sqrt{\frac{2\sqrt[32]{1+x^2}-\sqrt{\sqrt{\sqrt{\sqrt{2+2\sqrt{1+x^2}}+2\sqrt[4]{1+x^2}}+2\sqrt[8]{1+x^2}}+2\sqrt[16]{1+x^2}}}{2\sqrt[32]{1+x^2}+\sqrt{\sqrt{\sqrt{\sqrt{2+2\sqrt{1+x^2}}+2\sqrt[4]{1+x^2}}+2\sqrt[8]{1+x^2}}+2\sqrt[16]{1+x^2}}}},\quad x\le 0 \tag{1.13.6 b}$$

...

(1.13.7)

$$\tan\left(\frac{1}{2^n}\tan^{-1}(x)\right)=\sqrt{\frac{2\sqrt[2^n]{1+x^2}-\sqrt{\sqrt{\sqrt{2+2\sqrt{1+x^2}}+2\sqrt[4]{1+x^2}}+\ldots+2^{2^{n-1}}\sqrt{1+x^2}}}{2\sqrt[2^n]{1+x^2}+\sqrt{\sqrt{\sqrt{2+2\sqrt{1+x^2}}+2\sqrt[4]{1+x^2}}+\ldots+2^{2^{n-1}}\sqrt{1+x^2}}}},\quad x\geq 0 \qquad (1.13.7\,a)$$

$$\tan\left(\frac{1}{2^n}\tan^{-1}(x)\right)=-\sqrt{\frac{2\sqrt[2^n]{1+x^2}-\sqrt{\sqrt{\sqrt{2+2\sqrt{1+x^2}}+2\sqrt[4]{1+x^2}}+\ldots+2^{2^{n-1}}\sqrt{1+x^2}}}{2\sqrt[2^n]{1+x^2}+\sqrt{\sqrt{\sqrt{2+2\sqrt{1+x^2}}+2\sqrt[4]{1+x^2}}+\ldots+2^{2^{n-1}}\sqrt{1+x^2}}}},\quad x\leq 0 \qquad (1.13.7\,b)$$

Note: We can express both negative and positive values of expressions (1.13.2) to (1.13.7b) by using the absolute value of x, which is defined as

- |x| = x when x ≥ 0
- |x| = -x when x < 0.

If we set x = a/b, then (1.13.7a) and (1.13.7b) become

(1.13.8)

$$\tan\left(\frac{1}{2^n}\tan^{-1}\left(\frac{a}{b}\right)\right)=\left(\frac{2\left(a^2+b^2\right)^{1/2^n}-\left(\left(\left(2a+2\left(a^2+b^2\right)^{1/2}\right)+2\left(a^2+b^2\right)^{1/4}\right)^{1/2}\ldots+2\left(a^2+b^2\right)^{1/2^{n-1}}\right)^{1/2}}{2\left(a^2+b^2\right)^{1/2^n}+\left(\left(\left(2a+2\left(a^2+b^2\right)^{1/2}\right)+2\left(a^2+b^2\right)^{1/4}\right)^{1/2}\ldots+2\left(a^2+b^2\right)^{1/2^{n-1}}\right)^{1/2}}\right)^{1/2},\quad a\geq b \quad (1.13.8\,a)$$

$$\tan\left(\frac{1}{2^n}\tan^{-1}\left(\frac{a}{b}\right)\right)=-\left(\frac{2\left(a^2+b^2\right)^{1/2^n}-\left(\left(\left(2a+2\left(a^2+b^2\right)^{1/2}\right)+2\left(a^2+b^2\right)^{1/4}\right)^{1/2}\ldots+2\left(a^2+b^2\right)^{1/2^{n-1}}\right)^{1/2}}{2\left(a^2+b^2\right)^{1/2^n}+\left(\left(\left(2a+2\left(a^2+b^2\right)^{1/2}\right)+2\left(a^2+b^2\right)^{1/4}\right)^{1/2}\ldots+2\left(a^2+b^2\right)^{1/2^{n-1}}\right)^{1/2}}\right)^{1/2},\quad a\leq b \quad (1.13.8\,b)$$

2. Nested Radical Generated by Composition tanh[(1/2ⁿ)tanh⁻¹(x)]

We establish the nested radical expressions associated with the hyperbolic tangent function and inverse hyperbolic tangent function in the form:

$$\tanh\left(\frac{1}{2^n}\tanh^{-1}(x)\right),\quad \text{where } |x|<1,\text{ and } n\in N=\{1,2,3,\ldots\}. \qquad (1.13.9)$$

By applying the identities tan(ix) = itan(x) and tan⁻¹(ix) = itanh⁻¹(x), we easily transform the identities from (1.13.2) to (1.13.18) of Part 1 to the form of hyperbolic tangent (1.13.9). Below, we present a compilation of the derived identities.

$$\tanh\left(\frac{1}{2}\tanh^{-1}(x)\right)=\frac{\sinh\left(\frac{1}{2}\tanh^{-1}(x)\right)}{\cosh\left(\frac{1}{2}\tanh^{-1}(x)\right)}$$

$$=\begin{cases}\sqrt{\dfrac{1-\sqrt{1-x^2}}{1+\sqrt{1-x^2}}} & 0\le x<1\\[4ex] -\sqrt{\dfrac{1-\sqrt{1-x^2}}{1+\sqrt{1-x^2}}} & -1<x<0,\end{cases} \qquad (1.13.10)$$

$$\tanh\left(\frac{1}{4}\tanh^{-1}(x)\right)=\pm\sqrt{\frac{\sqrt{2+2\sqrt{1-x^2}}-2\sqrt[4]{1-x^2}}{\sqrt{2+2\sqrt{1-x^2}}+2\sqrt[4]{1-x^2}}}, \qquad (1.13.11)$$

$$\tanh\left(\frac{1}{8}\tanh^{-1}(x)\right)=\pm\sqrt{\frac{\sqrt{\sqrt{2+2\sqrt{1-x^2}}+2\sqrt[4]{1-x^2}}-2\sqrt[8]{1-x^2}}{\sqrt{\sqrt{2+2\sqrt{1-x^2}}+2\sqrt[4]{1-x^2}}+2\sqrt[8]{1-x^2}}}, \qquad (1.13.12)$$

$$\tanh\left(\frac{1}{16}\tanh^{-1}(x)\right)=\pm\sqrt{\frac{\sqrt{\sqrt{\sqrt{2+\sqrt{1-x^2}}+2\sqrt[4]{1-x^2}}+2\sqrt[8]{1-x^2}}-2\sqrt[16]{1-x^2}}{\sqrt{\sqrt{\sqrt{2+2\sqrt{1-x^2}}+2\sqrt[4]{1-x^2}}+2\sqrt[8]{1-x^2}}+2\sqrt[16]{1-x^2}}}, \qquad (1.12.13)$$

$$\tanh\left(\frac{1}{32}\tanh^{-1}(x)\right)=\pm\sqrt{\frac{\sqrt{\sqrt{\sqrt{\sqrt{2+2\sqrt{1-x^2}}+2\sqrt[4]{1-x^2}}+2\sqrt[8]{1-x^2}}+2\sqrt[16]{1-x^2}}-2\sqrt[32]{1-x^2}}{\sqrt{\sqrt{\sqrt{\sqrt{2+2\sqrt{1-x^2}}+2\sqrt[4]{1-x^2}}+2\sqrt[8]{1-x^2}}+2\sqrt[16]{1-x^2}}+2\sqrt[32]{1-x^2}}}, \qquad (1.13.14)$$

...

$$\tanh\left(\frac{1}{2^n}\tanh^{-1}(x)\right)=\pm\sqrt{\frac{\sqrt{\sqrt{\sqrt{2+2\sqrt{1-x^2}}+2\sqrt[4]{1-x^2}}+2\sqrt[8]{1-x^2}}+\ldots 2^{2^{n-1}}\sqrt{1-x^2}-2^{2^n}\sqrt{1-x^2}}{\sqrt{\sqrt{\sqrt{2+2\sqrt{1-x^2}}+2\sqrt[4]{1-x^2}}+2\sqrt[8]{1-x^2}}+\ldots 2^{2^{n-1}}\sqrt{1-x^2}+2^{2^n}\sqrt{1-x^2}}}. \qquad (1.13.15)$$

All expressions, (1.13.10) – (1.13.15), are positive when 0≤x<1, and they are negative when -1<x<0.

Moreover, if we set x = b/a, where a and b are real numbers, then (1.13.15) becomes

$$\tanh\left(\frac{1}{2^n}\tanh^{-1}\left(\frac{a}{b}\right)\right)=\pm\left(\frac{\left(\left(\left(2a+2\left(a^2-b^2\right)^{1/2}\right)+2\left(a^2-b^2\right)^{1/4}\right)^{1/2}\cdots+2\left(a^2-b^2\right)^{1/2^{n-1}}\right)^{1/2}-2\left(a^2-b^2\right)^{1/2^n}}{\left(\left(\left(2a+2\left(a^2-b^2\right)^{1/2}\right)+2\left(a^2-b^2\right)^{1/4}\right)^{1/2}\cdots+2\left(a^2-b^2\right)^{1/2^{n-1}}\right)^{1/2}+2\left(a^2-b^2\right)^{1/2^n}}\right)^{1/2}, \quad (1.13.16)$$

The expression (1.13.16) is positive when 0≤a<b, and it is negative when -b<a<0.

3. Establish Algebraic Identities

We establish algebraic identities based on the findings presented in Part 2 and the definition of hyperbolic tangent and inverse hyperbolic tangent as follows:

- The inverse hyperbolic tangent is defined as

$$\tanh^{-1}(x)=\frac{1}{2}\ln\left(\frac{1+x}{1-x}\right) \quad -1<x<1. \tag{1.13.17}$$

- Let n be a real number. Then we have

$$\tanh\left(\frac{1}{n}\tanh^{-1}(x)\right)=\tanh\left(\frac{1}{2n}\ln\left(\frac{1+x}{1-x}\right)\right) \tag{1.13.17}$$

$$=\frac{\left(\frac{1+x}{1-x}\right)^{1/2n}-\left(\frac{1+x}{1-x}\right)^{-1/2n}}{\left(\frac{1+x}{1-x}\right)^{1/2n}+\left(\frac{1+x}{1-x}\right)^{-1/2n}} \qquad \text{Using } \left(\frac{1+x}{1-x}\right)^{-1/2n}=\left(\frac{1-x}{1+x}\right)^{1/2n}$$

$$=\frac{\left(\frac{1+x}{1-x}\right)^{1/2n}-\left(\frac{1-x}{1+x}\right)^{1/2n}}{\left(\frac{1+x}{1-x}\right)^{1/2n}+\left(\frac{1-x}{1+x}\right)^{1/2n}} \qquad \text{Multiply numerator and denominator by } (1-x)^{1/2n}(1+x)^{1/2n}$$

$$=\frac{(1+x)^{2/2n}-(1-x)^{2/2n}}{(1+x)^{2/2n}+(1-x)^{2/2n}}$$

$$=\frac{(1+x)^{1/n}-(1-x)^{1/n}}{(1+x)^{1/n}+(1-x)^{1/n}}$$

$$\tanh\left(\frac{1}{n}\tanh^{-1}(x)\right)=\frac{\sqrt[n]{1+x}-\sqrt[n]{1-x}}{\sqrt[n]{1+x}+\sqrt[n]{1-x}}. \tag{1.13.17a}$$

Therefore, by substituting n with 2^n, we derive the desired general formula for the composite of hyperbolic tangent with inverse hyperbolic tangent with $x \in (-1,1)$:

$$\tanh\left(\frac{1}{2^n}\tanh^{-1}(x)\right) = \frac{\sqrt[2^n]{1+x} - \sqrt[2^n]{1-x}}{\sqrt[2^n]{1+x} + \sqrt[2^n]{1-x}} \qquad (1.13.17\,b)$$

Now, by combining (1.13.17b) with each identity ranging from (1.13.10) to (1.13.15) for n = 1, 2, ..., n, we obtain the following subsequent identities:

$$\frac{\sqrt{1+x}-\sqrt{1-x}}{\sqrt{1+x}+\sqrt{1-x}} = \pm\sqrt{\frac{1-\sqrt{1-x^2}}{1+\sqrt{1-x^2}}}, \qquad (1.13.18)$$

$$\frac{\sqrt[4]{1+x}-\sqrt[4]{1-x}}{\sqrt[4]{1+x}+\sqrt[4]{1-x}} = \pm\sqrt{\frac{\sqrt{2\sqrt{1+x^2}}-\sqrt{1+\sqrt{1+x^2}}}{\sqrt{2\sqrt{1+x^2}}+\sqrt{1+\sqrt{1+x^2}}}}, \qquad (1.13.19)$$

$$\frac{\sqrt[8]{1+x}-\sqrt[8]{1-x}}{\sqrt[8]{1+x}+\sqrt[8]{1-x}} = \pm\sqrt{\frac{2\sqrt[8]{1+x^2}-\sqrt{\sqrt{2+2\sqrt{1+x^2}}+2\sqrt[4]{1+x^2}}}{2\sqrt[8]{1+x^2}+\sqrt{\sqrt{2+2\sqrt{1+x^2}}+2\sqrt[4]{1+x^2}}}}, \qquad (1.13.20)$$

$$\frac{\sqrt[16]{1+x}-\sqrt[16]{1-x}}{\sqrt[16]{1+x}+\sqrt[16]{1-x}} = \pm\sqrt{\frac{2\sqrt[16]{1+x^2}-\sqrt{\sqrt{\sqrt{2+2\sqrt{1+x^2}}+2\sqrt[4]{1+x^2}}+2\sqrt[8]{1+x^2}}}{2\sqrt[16]{1+x^2}+\sqrt{\sqrt{\sqrt{2+2\sqrt{1+x^2}}+2\sqrt[4]{1+x^2}}+2\sqrt[8]{1+x^2}}}}, \qquad (1.13.21)$$

$$\frac{\sqrt[32]{1+x}-\sqrt[32]{1-x}}{\sqrt[32]{1+x}+\sqrt[32]{1-x}} = \pm\sqrt{\frac{2\sqrt[32]{1+x^2}-\sqrt{\sqrt{\sqrt{\sqrt{2+2\sqrt{1+x^2}}+2\sqrt[4]{1+x^2}}+2\sqrt[8]{1+x^2}}+2\sqrt[16]{1+x^2}}}{2\sqrt[32]{1+x^2}+\sqrt{\sqrt{\sqrt{\sqrt{2+2\sqrt{1+x^2}}+2\sqrt[4]{1+x^2}}+2\sqrt[8]{1+x^2}}+2\sqrt[16]{1+x^2}}}}, \qquad (1.13.22)$$

$$\frac{\sqrt[2^n]{1+x}-\sqrt[2^n]{1-x}}{\sqrt[2^n]{1+x}+\sqrt[2^n]{1-x}} = \pm\sqrt{\frac{2\sqrt[2^n]{1+x^2}-\sqrt{\sqrt{\sqrt{2+2\sqrt{1+x^2}}+2\sqrt[4]{1+x^2}}+2\sqrt[8]{1+x^2}+...2^{2^{n-1}}\sqrt{1+x^2}}}{2\sqrt[2^n]{1+x^2}+\sqrt{\sqrt{\sqrt{2+2\sqrt{1+x^2}}+2\sqrt[4]{1+x^2}}+2\sqrt[8]{1+x^2}+...2^{2^{n-1}}\sqrt{1+x^2}}}}, \qquad (1.13.23)$$

where the positive sign is used when $0 \le x \le 1$, and the negative sign is used when $-1 \le x < 0$. *Importantly, the identities remain valid for all x when they are complex numbers.*

Section 1-14. Nested Radical Generated by Composition sin[(1/2ⁿ)sin⁻¹(x)], and Composition sinh[(1/2ⁿ)sinh⁻¹(x)]

1. Nested Radical Generated by Composition sin[(1/2ⁿ)sin⁻¹(x)]

We develop the nested radical expressions associated with the sine function and inverse sine function with the structure:

$$\sin\left(\frac{1}{2^n}\sin^{-1}(x)\right), -1\leq x\leq 1, \quad n\in \mathbb{N}. \tag{1.14.1}$$

A. Half Angle Sine Formula

To derive half angle sine formula, we begin by squaring both sides of (1.8.3), yielding

$$\cos^2\left(\frac{1}{2}\sin^{-1}(x)\right)=\frac{1}{2}\left(1+\sqrt{1-x^2}\right).$$

Next, employing the Pythagorean trigonometric identity to the left-hand side of the expression and simplifying further, we get

$$\sin^2\left(\frac{1}{2}\sin^{-1}(x)\right)=\frac{1}{2}\left(1-\sqrt{1-x^2}\right).$$

Taking the square root of both sides leads us to the final result,

$$\sin\left(\frac{1}{2}\sin^{-1}(x)\right)=\begin{cases}\frac{1}{2}\sqrt{2-2\sqrt{1-x^2}} & \text{for} \quad 0\leq x\leq 1 \\[2mm] -\frac{1}{2}\sqrt{2-2\sqrt{1-x^2}} & \text{for} \quad -1\leq x<0.\end{cases} \tag{1.14.2}$$

The expression on the left-hand side of (1.14.2) is defined -1≤x≤1 due to the restriction on the domain of the inverse sine while the domains on the right-hand side of (1.14.2) are determined as follows:

$$\begin{cases}2-2\sqrt{1-x^2}\geq 0 \\ \text{and} \quad 1-x^2\geq 0\end{cases} \Rightarrow \begin{cases}\sqrt{1-x^2}\leq 1 \\ \text{and} -1\leq x\leq 1\end{cases} \Rightarrow \begin{cases}x^2\geq 0 \\ \text{and} -1\leq x\leq 1\end{cases} \Rightarrow \begin{cases}0\leq x\leq 1 \\ \text{or} -1\leq x<0.\end{cases}$$

B. Quarter Angle Sine Formula

The half angle sine formula emerges as a consequence of applying (1.14.1) with n = 2. To derive this result, we begin by squaring both sides of (1.8.4), yielding

$$\cos^2\left(\frac{1}{4}\sin^{-1}(x)\right)=\frac{1}{4}\left(2+\sqrt{2+2\sqrt{1-x^2}}\right).$$

Applying Pythagorean trigonometric identity and deducing that

$$\sin^2\left(\frac{1}{4}\sin^{-1}(x)\right)=\frac{1}{4}\left(2-\sqrt{2+2\sqrt{1-x^2}}\right).$$

Taking the square root of both sides, gives the final result

$$\sin\left(\frac{1}{4}\sin^{-1}(x)\right)=\begin{cases}\dfrac{1}{2}\sqrt{2-\sqrt{2+2\sqrt{1-x^2}}} & 0\le x\le 1 \\[2ex] -\dfrac{1}{2}\sqrt{2-\sqrt{2+2\sqrt{1-x^2}}} & -1\le x<0.\end{cases} \tag{1.14.3}$$

Following the established procedure, we proceed to deduce the subsequent identities of (1.14.1) when n = 3, 4, 5,...n, as presented below:

$$\sin\left(\frac{1}{8}\sin^{-1}x\right)=\pm\frac{1}{2}\sqrt{2-\sqrt{2+\sqrt{2+2\sqrt{1-x^2}}}}, \tag{1.14.4}$$

$$\sin\left(\frac{1}{16}\sin^{-1}x\right)=\pm\frac{1}{2}\sqrt{2-\sqrt{2+\sqrt{2+\sqrt{2+2\sqrt{1-x^2}}}}}, \tag{1.14.5}$$

$$\sin\left(\frac{1}{32}\sin^{-1}x\right)=\pm\frac{1}{2}\sqrt{2-\sqrt{2+\sqrt{2+\sqrt{2+\sqrt{2+2\sqrt{1-x^2}}}}}}, \tag{1.14.6}$$

$$\sin\left(\frac{1}{64}\sin^{-1}x\right)=\pm\frac{1}{2}\sqrt{2-2\sqrt{2+\sqrt{2+\sqrt{2+\sqrt{2+2\sqrt{1-x^2}}}}}}, \tag{1.14.7}$$

$$\sin\left(\frac{1}{128}\sin^{-1}x\right)=\pm\frac{1}{2}\sqrt{2-\sqrt{2+\sqrt{2+\sqrt{2+\sqrt{2+\sqrt{2+2\sqrt{1-x^2}}}}}}}, \tag{1.14.8}$$

...

$$\sin\left(\frac{1}{2^n}\sin^{-1}(x)\right)=\pm\frac{1}{2}\sqrt{2-\sqrt{2+\ldots\sqrt{2+\sqrt{2+\sqrt{2+\sqrt{2+\sqrt{2+2\sqrt{1-x^2}}}}}}}}. \tag{1.14.9}$$

In each of these equations spanning from (1.14.4) to (1.14.9), the expressions on the right-hand side exhibit consistent behavior: they are positive for $0 \le x \le 1$ and negative for $-1 \le x < 0$.

2. Nested Radical Generated by Composition sinh[(1/2ⁿ)sinh⁻¹(x)]

We construct the composition of the hyperbolic sine with its inverse in the form

$$\sinh\left(\frac{1}{2^n}\sinh^{-1}(x)\right), \quad x\in\mathbb{R}, \quad n\in\mathbb{N}=\{1,2,3,\ldots\}. \tag{1.14.10}$$

A. Half Hyperbolic Sine Formula

By applying the identities $\sin(ix) = i\sinh(x)$ and $\sin^{-1}(ix) = i\sinh^{-1}(x)$ in (1.14.2), we deduce that

$$i\sinh\left(\frac{1}{2}\sinh^{-1}(x)\right)=\begin{cases}\dfrac{1}{2}\sqrt{2-2\sqrt{1+x^2}}\\[2mm]-\dfrac{1}{2}\sqrt{2-2\sqrt{1+x^2}}.\end{cases}$$

Because $2(1+x^2)$ is greater than 2 for all x, the expression can be rewritten as

$$i\sinh\left(\frac{1}{2}\sinh^{-1}(x)\right)=\begin{cases}\dfrac{i}{2}\sqrt{2\sqrt{1+x^2}-2}\\[2mm]-\dfrac{i}{2}\sqrt{2\sqrt{1+x^2}-2}.\end{cases}$$

Then, by simplifying both sides with i, we obtain the formula

$$\sinh\left(\frac{1}{2}\sinh^{-1}(x)\right)=\begin{cases}\dfrac{1}{2}\sqrt{2\sqrt{1+x^2}-2} & \text{for } x\ge 0\\[2mm]-\dfrac{1}{2}\sqrt{2\sqrt{1+x^2}-2} & \text{for } x<0,\end{cases} \tag{1.14.11}$$

which is the result from the setting in (1.14.10) when n = 1.

B. Quarter Hyperbolic Sine Formula

Similarly, by substituting the identities sin(ix) = isinh(x) and sin^{-1}(ix) = isinh^{-1}(x) into (1.14.3), we deduce that

$$\text{isinh}\left(\frac{1}{4}\sinh^{-1}(x)\right)=\begin{cases}\frac{1}{2}\sqrt{2-\sqrt{2+2\sqrt{1+x^2}}}\\[2mm]-\frac{1}{2}\sqrt{2-\sqrt{2+2\sqrt{1+x^2}}}\end{cases}$$

$$=\begin{cases}\frac{i}{2}\sqrt{\sqrt{2+2\sqrt{1+x^2}}-2}\\[2mm]-\frac{i}{2}\sqrt{\sqrt{2+2\sqrt{1+x^2}}-2}.\end{cases}$$

Upon simplifying both sides with i, we obtain

$$\sinh\left(\frac{1}{4}\sinh^{-1}(x)\right)=\begin{cases}\frac{1}{2}\sqrt{\sqrt{2+2\sqrt{1+x^2}}-2},& x\geq 0\\[2mm]-\frac{1}{2}\sqrt{\sqrt{2+2\sqrt{1+x^2}}-2},& x<0,\end{cases}\qquad(1.14.12)$$

which is the result from the setting in (1.14.10) when n = 2.

In the same manner, employing the identical approach yields other subsequent identities of (1.14.10) when n = 3, 4, 5,...n, as presented below:

$$\sinh\left(\frac{1}{8}\sinh^{-1}(x)\right)=\pm\frac{1}{2}\sqrt{\sqrt{2+\sqrt{2+2\sqrt{1+x^2}}}-2},\qquad(1.14.13)$$

$$\sinh\left(\frac{1}{16}\sinh^{-1}(x)\right)=\pm\frac{1}{2}\sqrt{\sqrt{2+\sqrt{2+\sqrt{2+2\sqrt{1-x^2}}}}+2},\qquad(1.14.14)$$

$$\sinh\left(\frac{1}{32}\sinh^{-1}(x)\right)=\pm\frac{1}{2}\sqrt{\sqrt{2+\sqrt{2+\sqrt{2+\sqrt{2+2\sqrt{1+x^2}}}}}-2},\qquad(1.14.15)$$

$$\sinh\left(\frac{1}{64}\sinh^{-1}(x)\right)=\pm\frac{1}{2}\sqrt{2+\sqrt{2+\sqrt{2+\sqrt{2+\sqrt{2+2\sqrt{1+x^2}}}}}}-2,$$

(1.14.16)

$$\sinh\left(\frac{1}{128}\sinh^{-1}(x)\right)=\pm\frac{1}{2}\sqrt{2+\sqrt{2+\sqrt{2+\sqrt{2+\sqrt{2+\sqrt{2+2\sqrt{1+x^2}}}}}}}-2,$$

(1.14.17)

$$\sinh\left(\frac{1}{2^n}\sinh^{-1}(x)\right)=\pm\frac{1}{2}\sqrt{2+...\sqrt{2+\sqrt{2+\sqrt{2+\sqrt{2+\sqrt{2+2\sqrt{1+x^2}}}}}}}-2,$$

(1.14.18)

or displaying (1.14.18) in "another form style",

$$\sinh\left(\frac{1}{2^n}\sinh^{-1}(x)\right)=\pm\frac{1}{2}\sqrt{\sqrt{\sqrt{\sqrt{\sqrt{\sqrt{2\sqrt{1+x^2}+2}+2}+2}+2}+...2}-2}.$$

(1.14.18)

In each of these equations spanning from (1.14.13) to (1.14.18), the positive sign is used when x is greater than or equal to 0, and the negative sign is used when x is less than or equal to 0.

Section 2. Some Special Nested Radicals

As we mentioned in Part 3 of Section 1-2, we have yet to discover a nested radical for odd numbers that consists of only two terms on its left hand side while ensuring that the expression on its right-hand side is free from the complex unit 'i' and doesn't duplicate the original two terms from the left hand side. While conducting online research, we came across an elegant cubic radical identity originally unearthed by Srinivasa Ramanujan who gave the radical with three terms as shown below:

$$\frac{1}{3}\left(\sqrt[3]{(4m+n)^2}+\sqrt[3]{4(m-2n)(4m+n)}-\sqrt[3]{2(m-2n)^2}\right)=\sqrt{m\sqrt[3]{4m-8n}+n\sqrt[3]{4m+n}},$$

where m and n are arbitrary.

This distinctive formula originates from Ramanujan's Notebooks, Part IV [3], meticulously compiled by Bruce C. Berndt. Fortunately, we also explore similar formulas for quintic, septic, and nonic nested radicals that share the same patterns as those Ramanujan discovered. We will present the findings in the following sub-sections.

Section 2-1. Quintic Nested Radical (Fifth Root)

We present the finding of quintic nested radical expression through transformation method. Let m and n be arbitrary and we observe that:

$$\left(m^2-n^2+mn\right)^3=m\left(3n^5+m^5\right)+n\left(3m^5-n^5\right)-5m^3n^3. \tag{2.1.1}$$

Setting

$$3n^5+m^5=a$$

and

$$3m^5-n^5=b,$$

where a and b are real numbers.

Upon solving for m and n in terms of a and b, we derive:

$$m=\left(\frac{a+3b}{10}\right)^{1/5}$$

and

$$n=\left(\frac{3a-b}{10}\right)^{1/5}.$$

By substituting the expressions for m and n into (2.1.1), we get

$$\left(\left(\frac{a+3b}{10}\right)^{2/5}-\left(\frac{3a-b}{10}\right)^{2/5}+\left(\frac{a+3b}{10}\right)^{1/5}\left(\frac{3a-b}{10}\right)^{1/5}\right)^{3}=\left(\frac{a+3b}{10}\right)^{1/5}a+\left(\frac{3a-b}{10}\right)^{1/5}b-5\left(\frac{a+3b}{10}\right)^{3/5}\left(\frac{3a-b}{10}\right)^{3/5}$$

Simplifying further, we obtain nested quintic radical (fifth root),

$$\sqrt[5]{(a+3b)^2}-\sqrt[5]{(3a-b)^2}+\sqrt[5]{3(a^2-b^2)+8ab}=\sqrt[3]{10a\sqrt[5]{a+3b}+10b\sqrt[5]{3a-b}-5\sqrt[5]{[3(a^2-b^2)+8ab]^3}}. \quad (2.1.2)$$

Section 2-2. Septic Nested Radical (Seventh Root)

Similar to section 2-1, we observe another algebraic expression,

$$\left(m^2-3n^2+3mn\right)^4=m\left(m^7-324n^7\right)+3n\left(27n^7+4m^7\right)+42m^2n^2\left(9n^4+m^4\right)-189m^4n^4 \quad (2.2.1).$$

By setting

$$m^7-324n^7=a$$

and

$$27n^7+4m^7=b,$$

where a and b are real numbers.

Solve for m and n in terms of a and b, we obtain

$$m=\left(\frac{a+12b}{49}\right)^{1/7}$$

and

$$n=\left(\frac{b-4a}{1323}\right)^{1/7}.$$

Substituting the expressions m and n into (2.2.1) yields the septic radical (seventh root):

$$\sqrt[7]{\left(\frac{a+12b}{49}\right)^2}-3\sqrt[7]{\left(\frac{b-4a}{1323}\right)^2}+3\sqrt[7]{\left(\frac{a+12b}{49}\right)\left(\frac{b-4a}{1323}\right)}= \quad (2.2.2)$$

$$\sqrt[4]{a\sqrt[7]{\frac{a+12b}{49}}+3b\sqrt[7]{\frac{b-4a}{1323+42}}+\sqrt[7]{\left(\frac{a+12b}{49}\right)^{2}\left(\frac{y-4a}{1323}\right)^{2}}\left(9\sqrt[7]{\left(\frac{b-4a}{1323}\right)^{4}}+\sqrt[7]{\left(\frac{a+12b}{49}\right)^{4}}\right)-189\sqrt[7]{\left(\frac{b-4a}{1323}\right)\left(\frac{a+12b}{49}\right)^{4}}}.$$

Section 2-3. Nonic Nested Radical (Ninth Root)

Let m and n be real numbers, we observe that

$$\left(m^{2}-2n^{2}+mn\right)^{5}=m\left(m^{9}+80n^{9}\right)+n\left(5m^{9}-32n^{9}\right)-30m^{3}n^{3}\left(4n^{4}+m^{4}\right)+15m^{4}n^{4}\left(2n^{2}-m^{2}\right)+81m^{5}n^{5}. \qquad (2.3.1)$$

By setting $m^{9}+80\,n^{9}=a$ and $5\,m^{9}-32\,n^{9}=b$,

where a and b are real.

Solving for m and n in terms of a and b, we obtain

$$m=\left(\frac{2a+5b}{27}\right)^{1/9}$$

and

$$n=\left(\frac{5a-b}{432}\right)^{1/9}.$$

By substituting the expressions m and n into (2.3.1), which gives the nonic nested radical in this form,

$$\left(\sqrt[9]{\left(\frac{2a+5b}{27}\right)^{2}}-2\sqrt[9]{\left(\frac{5a-b}{432}\right)^{2}}+\sqrt[9]{\frac{2a+5b}{27}}\sqrt[9]{\frac{5a-b}{432}}\right)^{5}= \qquad (2.3.2)$$

$$a\sqrt[9]{\frac{2a+5b}{27}}+b\sqrt[9]{\frac{5a-b}{432}}-30\sqrt[3]{\frac{2a+5b}{27}}\sqrt[3]{\frac{5a-b}{432}}\left(4\sqrt[9]{\left(\frac{5a-b}{432}\right)^{4}}+\sqrt[9]{\left(\frac{2a+5b}{27}\right)^{4}}\right)$$

$$+15\sqrt[9]{\left(\frac{2a+5b}{27}\right)^{4}}\sqrt[9]{\left(\frac{5a-b}{432}\right)^{4}}\left(2\sqrt[9]{\left(\frac{5a-b}{432}\right)^{2}}-\sqrt[9]{\left(\frac{2a+5b}{27}\right)^{2}}\right)+81\sqrt[9]{\left(\frac{2a+5b}{27}\right)^{5}}\sqrt[9]{\left(\frac{5a-b}{432}\right)^{5}}$$

Taking the fifth root of both its sides gives

$$\sqrt[9]{\left(\frac{2a+5b}{27}\right)^{2}}-2\sqrt[9]{\left(\frac{5a-b}{432}\right)^{2}}+\sqrt[9]{\frac{2a+5b}{27}}\sqrt[9]{\frac{5a-b}{432}}= \qquad (2.3.3)$$

46

$$\sqrt[5]{\left(\begin{matrix} a\sqrt[9]{\dfrac{2a+5b}{27}}+b\sqrt[9]{\dfrac{5a-b}{432}}-30\sqrt[3]{\dfrac{2a+5b}{27}}\sqrt[3]{\dfrac{5a-b}{432}}\left(4\sqrt[9]{\left(\dfrac{5a-b}{432}\right)^4}+\sqrt[9]{\left(\dfrac{2a+5b}{27}\right)^4}\right) \\ +15\sqrt[9]{\left(\dfrac{2a+5b}{27}\right)^4}\sqrt[9]{\left(\dfrac{5a-b}{432}\right)^4}\left(2\sqrt[9]{\left(\dfrac{5a-b}{432}\right)^2}-\sqrt[9]{\left(\dfrac{2a+5b}{27}\right)^2}\right)+81\sqrt[9]{\left(\dfrac{2a+5b}{27}\right)^5}\sqrt[9]{\left(\dfrac{5a-b}{432}\right)^5} \end{matrix}\right)}.$$

Section 3. Complex Radical Subtraction and Addition Formulas

In this section, we extend the scope of the formulas of radical subtraction and addition to encompass a form of complex radicals.

Section 3-1. General Subtraction and Addition Formulas

We present a method to express a form of the complex radical in other several forms. Before we do it, let us recapture what we already know. The concept of representing complex numbers in the form [4]:

$$a+ib=r\left(\cos\theta+i\sin\theta\right), \quad \text{where} \quad r=\sqrt{a^2+b^2}, \text{and } \theta=\tan^{-1}\left(\frac{a}{b}\right). \tag{3.1}$$

Formula (3.1) is known as Euler's formula and is used to represent complex numbers in the polar form, in which:

- a and b are real numbers.
- a represents the real part of the complex number.
- b represents the imaginary part of the complex number.
- r represents the magnitude (or modulus) of the complex number.
- θ represents the argument (or angle) of the complex number.

Please note that while it is possible to simplify formula (3.1) using the cis function for clarity, we have chosen not to use this approach in the subsequent section. This decision is to ensure that readers can closely follow the detailed overview of each expression as we progress.

Now, let n be a real number, raise each side of (3.1) to the power 1/n, and apply de Moivre's theorem [5] to its right-hand side, we obtain the general addition complex n^{th} root,

$$(a+ib)^{1/n}=\left(a^2+b^2\right)^{1/(2n)}\left[\cos\left(\frac{1}{n}\tan^{-1}\left(\frac{b}{a}\right)\right)+i\sin\left(\frac{1}{n}\tan^{-1}\left(\frac{b}{a}\right)\right)\right]. \tag{3.2}$$

By replacing b with -b, which gives a form of subtraction complex n^{th} root,

$$(a-ib)^{1/n} = (a^2+b^2)^{1/(2n)}\left[\cos\left(\frac{1}{n}\tan^{-1}\left(\frac{b}{a}\right)\right) - i\sin\left(\frac{1}{n}\tan^{-1}\left(\frac{b}{a}\right)\right)\right].$$

(3.3)

If we replace n with 2^n into both (3.2) and (3.3), we deduce that

$$(a\pm ib)^{1/2^n} = (a^2+b^2)^{1/2^{n+1}}\left[\cos\left(\frac{1}{2^n}\tan^{-1}\left(\frac{a}{b}\right)\right) \pm i\sin\left(\frac{1}{2^n}\tan^{-1}\left(\frac{b}{a}\right)\right)\right].$$

(3.3 a)

Next, we establish two formulas: Complex Radical Subtraction and Complex Radical Addition.

General Complex Radical Subtraction Formula

By subtracting (3.3) from (3.2), we obtain the subtraction formula,

$$(a+ib)^{1/n} - (a-ib)^{1/n} = 2i(a^2+b^2)^{1/2n}\sin\left(\frac{1}{n}\tan^{-1}\frac{b}{a}\right).$$

(3.3 b)

Formula (3.3b) illustrates an important result as two complex n^{th} roots are subtracted from each other, the result is a complex number.

General Complex Radical Addition Formula

Adding (3.2) and (3.3) yields the addition formula,

$$(a+ib)^{1/n} + (a-ib)^{1/n} = 2(a^2+b^2)^{1/2n}\cos\left(\frac{1}{n}\tan^{-1}\frac{b}{a}\right).$$

(3.3 c)

Formula (3.3c) demonstrates that the addition of two complex n^{th} roots yields a real number as the result.

Section 3-2. Complex Formulas for Adding and Subtracting 2^{nth} Radicals

We present formulas for adding and subtracting 2^{nth} radicals. By employing formula (3.3a) in conjunction with the cosine and sine formulas from Sections 1-10 and 1-12, we can derive a closed-form for the expressions of 2nd, 4th, 8th,..., and 2^{nth} radicals. We then subsequently establish formulas for their complex nested radical expressions in both subtraction and addition.

1. Square Root of (a±bi):

By substituting cosine from (1.10.3) and sine from (1.12.2) for into (3.3a) for n = 1, and then setting x = b/a, we derive the following expression:

$$\sqrt{a \pm ib} = \frac{1}{2}\sqrt[4]{a^2+b^2}\left(\sqrt{2+\frac{2a}{\sqrt{a^2+b^2}}} \pm i\sqrt{2-\frac{2a}{\sqrt{b^2+a^2}}}\right)$$

$$= \frac{\sqrt{2}}{2}\sqrt{\sqrt{a^2+b^2}+a} \pm i\frac{\sqrt{2}}{2}\sqrt{\sqrt{a^2+b^2}-a}$$

Therefore, the square root of (a + ib), when considering the positive sign, yields:

$$\sqrt{a+ib} = \frac{\sqrt{2}}{2}\sqrt{\sqrt{a^2+b^2}+a} + i\frac{\sqrt{2}}{2}\sqrt{\sqrt{a^2+b^2}-a}. \tag{3.4}$$

Similarly, by replacing b with -b, the square root of (a – ib) can be expressed as

$$\sqrt{a-ib} = \pm\frac{\sqrt{2}}{2}\left(\sqrt{\sqrt{a^2+b^2}+a} - i\sqrt{\sqrt{a^2+b^2}-a}\right), \tag{3.5}$$

$$\sqrt{a-ib} = \begin{cases} \dfrac{\sqrt{2}}{2}\left(\sqrt{\sqrt{a^2+b^2}+a} - i\sqrt{\sqrt{a^2+b^2}-a}\right), & a \geq b \tag{3.5a} \\[2mm] -\dfrac{\sqrt{2}}{2}\left(\sqrt{\sqrt{a^2+b^2}+a} - i\sqrt{\sqrt{a^2+b^2}-a}\right), & a < b \tag{3.5b} \end{cases}$$

Notice that *the identity remains valid when a and b are arbitrary complex numbers.*

Example. Use (3.5a) with a = 5, b = 2, we obtain a closed form of the square root of 5+2i,

$$\sqrt{5 \pm 2i} = \frac{\sqrt{2}}{2}\sqrt{\sqrt{29}+5} \pm i\frac{\sqrt{2}}{2}\sqrt{\sqrt{29}-5}.$$

Complex Square Root Addition Formula

The formula of complex square root addition is derived by adding (3.4) and (3.5) as follows:

$$\sqrt{a+ib} + \sqrt{a-ib} = \sqrt{2\sqrt{a^2+b^2}+a}. \tag{3.5c}$$

Example. By putting a = 7, b = 3 in (3.5c), the result is a real number,

$$\sqrt{7+3i} + \sqrt{7-3i} = \sqrt{2\sqrt{7^2+3^2}+7} = \sqrt{2\sqrt{58}+7}.$$

Complex Square Root Subtraction Formula

By subtracting (3.5) from (3.4), we obtain the subtraction formula,

$$\sqrt{a+ib}-\sqrt{a-ib}=i\cdot\sqrt{2\sqrt{a^2+b^2}-a}. \tag{3.5 d}$$

Example. Using (3.5d) with a = 7, b = 3 gives the result as a complex number,

$$\sqrt{7+3i}-\sqrt{7-3i}=i\sqrt{2\sqrt{7^2+3^2}+7}=i\sqrt{2\sqrt{58}-7}$$

2. Fourth Root of (a±bi):

Similarly, by substituting cosine from (1.10.4) and sine from (1.12.3) into (3.3a) for n = 2, and then setting x = b/a, we derive the following result:

$$\left(a\pm ib\right)^{1/4}=\frac{1}{2}\left(a^2+b^2\right)^{1/8}\left[\sqrt{2+\sqrt{2+\frac{2}{\sqrt{1+\left(\frac{b}{a}\right)^2}}}}\pm i\sqrt{2-\sqrt{2+\frac{2}{\sqrt{1+\left(\frac{b}{a}\right)^2}}}}\right]$$

By simplifying further, we deduce that

$$\sqrt[4]{a+ib}=\frac{1}{2}\left(\sqrt{2\sqrt[4]{a^2+b^2}+\sqrt{2\sqrt{a^2+b^2}+2a}}+i\sqrt{2\sqrt[4]{a^2+b^2}-\sqrt{2\sqrt{a^2+b^2}+2a}}\right) \tag{3.6}$$

and

$$\sqrt[4]{a-ib}=\begin{cases}\frac{1}{2}\left[\sqrt{2\sqrt[4]{a^2+b^2}+\sqrt{2\sqrt{a^2+b^2}+2a}}-i\sqrt{2\sqrt[4]{a^2+b^2}-\sqrt{2\sqrt{a^2+b^2}+2a}}\right], & a\geq b \quad (3.7\,a)\\[2em] -\frac{1}{2}\left[\sqrt{2\sqrt[4]{a^2+b^2}+\sqrt{2\sqrt{a^2+b^2}+2a}}-i\sqrt{2\sqrt[4]{a^2+b^2}-\sqrt{2\sqrt{a^2+b^2}+2a}}\right], & a<b. \quad (3.7\,b)\end{cases} \tag{3.7}$$

Complex Fourth Root Addition Formula

The complex fourth root addition formula is derived by adding (3.6) and (3.7a) as follows:

$$\sqrt[4]{a+ib}+\sqrt[4]{a-ib}=\sqrt{2\sqrt[4]{a^2+b^2}+\sqrt{2\sqrt{a^2+b^2}+2a}}.$$

Complex Fourth Root Subtraction Formula

Similar, the complex four root subtraction formula can be obtained by subtracting (3.7a) from (3.6):

$$\sqrt[4]{a+ib}-\sqrt[4]{a-ib}=i\sqrt{2\sqrt[4]{a^2+b^2}-\sqrt{2\sqrt{a^2+b^2}+2a}}$$

3. Eighth, 16th, 32th, ..., 2^{nth} Roots of (a±bi):

In the same manner, employing the identical approach yields other subsequent identities of (3.4) and (3.5) when n = 3, 4, 5,..., as presented below:

$$(a+ib)^{1/8}=\frac{1}{2}\left(a^2+b^2\right)^{1/16}\left[\sqrt{2+\sqrt{2+\sqrt{2+\frac{2}{\sqrt{1+\left(\frac{b}{a}\right)^2}}}}}+i\sqrt{2-\sqrt{2+\sqrt{2+\frac{2}{\sqrt{1+\left(\frac{b}{a}\right)^2}}}}}\right],\qquad(3.8)$$

$$(a-ib)^{1/8}=\pm\frac{1}{2}\left(a^2+b^2\right)^{1/16}\left[\sqrt{2+\sqrt{2+\sqrt{2+\frac{2}{\sqrt{1+\left(\frac{b}{a}\right)^2}}}}}-i\sqrt{2-\sqrt{2+\sqrt{2+\frac{2}{\sqrt{1+\left(\frac{b}{a}\right)^2}}}}}\right],\qquad(3.9)$$

$$(a+ib)^{1/8}+(a-ib)^{1/8}=\left(a^2+b^2\right)^{1/16}\sqrt{2+\sqrt{2+\sqrt{2+\frac{2}{\sqrt{1+\left(\frac{b}{a}\right)^2}}}}},$$

Adding (3.8) and (3.9) gives Complex 8th Root Addition Formula.

$$(a+ib)^{1/8}-(a-ib)^{1/8}=i\left(a^2+b^2\right)^{1/16}\sqrt{2-\sqrt{2+\sqrt{2+\frac{2}{\sqrt{1+\left(\frac{b}{a}\right)^2}}}}},$$

Subtracting (3.9) from (3.8) gives Complex 8th Root Subtraction Formula.

$$(a+ib)^{1/16}=\frac{1}{2}\left(a^2+b^2\right)^{1/32}\left[\sqrt{2+\sqrt{2+\sqrt{2+\sqrt{2+\frac{2}{\sqrt{1+\left(\frac{b}{a}\right)^2}}}}}}+i\sqrt{2-\sqrt{2+\sqrt{2+\sqrt{2+\frac{2}{\sqrt{1+\left(\frac{b}{a}\right)^2}}}}}}\right],\qquad(3.10)$$

51

$$(a-ib)^{1/16}=\pm\frac{1}{2}\left(a^2+b^2\right)^{1/32}\left[\sqrt{2+\sqrt{2+\sqrt{2+\sqrt{2+\frac{2}{\sqrt{1+(b/a)^2}}}}}}-i\sqrt{2-\sqrt{2+\sqrt{2+\sqrt{2+\frac{2}{\sqrt{1+(b/a)^2}}}}}}\right],\qquad(3.11)$$

$$(a+ib)^{1/16}+(a-ib)^{1/16}=\left(a^2+b^2\right)^{1/32}\sqrt{2+\sqrt{2+\sqrt{2+\sqrt{2+\frac{2}{\sqrt{1+(b/a)^2}}}}}},(3.12)$$

Adding (3.10) and (3.11) gives Complex 16th Root Addition Formula.

$$(a+ib)^{1/16}-(a-ib)^{1/16}=i\left(a^2+b^2\right)^{1/32}\sqrt{2-\sqrt{2+\sqrt{2+\sqrt{2+\frac{2}{\sqrt{1+(b/a)^2}}}}}},(3.13)$$

Subtracting (3.11) from (3.10) gives Complex 16th Root Subtraction Formula.

...

$$(a+ib)^{1/2^n}=\frac{1}{2}\left(a^2+b^2\right)^{1/2^{n+1}}\left[\sqrt{2+\sqrt{2+\sqrt{2+...\sqrt{2+\frac{2}{\sqrt{1+(b/a)^2}}}}}}+i\sqrt{2-\sqrt{2+\sqrt{2+...\sqrt{2+\frac{2}{\sqrt{1+(b/a)^2}}}}}}\right],\qquad(3.14)$$

$$(a-ib)^{1/2^n}=\pm\frac{1}{2}\left(a^2+b^2\right)^{1/2^{n+1}}\left[\sqrt{2+\sqrt{2+\sqrt{2+...\sqrt{2+\frac{2}{\sqrt{1+(b/a)^2}}}}}}-i\sqrt{2-\sqrt{2+\sqrt{2+...\sqrt{2+\frac{2}{\sqrt{1+(b/a)^2}}}}}}\right],\qquad(3.15)$$

$$(a+ib)^{1/2^n}+(a-ib)^{1/2^n}=\left(a^2+b^2\right)^{1/2^{n+1}}\sqrt{2+\sqrt{2+\sqrt{2+...\sqrt{2+\frac{2}{\sqrt{1+(b/a)^2}}}}}},(1.16)$$

Adding (3.14) and (3.15) gives Complex 2nth Root Addition Formula.

$$(a+ib)^{1/2^n}-(a-ib)^{1/2^n}=i\left(a^2+b^2\right)^{1/2^{n+1}}\sqrt{2-\sqrt{2+\sqrt{2+...\sqrt{2+\frac{2}{\sqrt{1+(b/a)^2}}}}}},(1.17)$$

Subtracting (3.15) from (3.14) gives Complex 2nth Root Subtraction Formula.

Formulas (3.14), (3.15), (3.16) and (3.17) comprises n layers of square roots, excluding the square root of $(1+b^2/a^2)$. The positive sign of identities (3.9), (3.11), and (3.13) is used when a ≥ b; otherwise, the negative sign is used.

Note that the algebraic expressions in this chapter remain valid even when a, b, m, n, A, or B are treated as arbitrary numbers, deviating from their initial assumption as real numbers.

Miscellaneous Examples

Example. Determine the radical expression for the half, quarter and 1/2[nth] of the lemniscate sine function [6], as given by

$$\text{sl}\left(\frac{1}{2}\text{arcsl}(x)\right) = \frac{\sin\left(\frac{1}{2}\sin^{-1}(x)\right)}{\cosh\left(\frac{1}{2}\sinh^{-1}(x)\right)} \tag{4.1}$$

and

$$\text{sl}\left(\frac{1}{4}\text{arcsl}(x)\right), \tag{4.2}$$

where sl is the lemniscate sine function and arcsl is the inverse function of sl for x within the interval [-1, 1].

Solution. By applying the following two formulas

$$\sin\left(\frac{1}{2}\sin^{-1}(x)\right) = \frac{1}{2}\sqrt{2 - 2\sqrt{1-x^2}}$$

and

$$\cosh\left(\frac{1}{2}\sinh^{-1}(x)\right) = \frac{1}{2}\sqrt{2 + 2\sqrt{1+x^2}}$$

in (4.1), which gives:

$$\text{sl}\left(\frac{1}{2}\text{arcsl}(x)\right) = \frac{\sin\left(\frac{1}{2}\sin^{-1}(x)\right)}{\cosh\left(\frac{1}{2}\sinh^{-1}(x)\right)}$$

$$= \frac{\dfrac{1}{2}\sqrt{2-2\sqrt{1-x^2}}}{\dfrac{1}{2}\sqrt{2+2\sqrt{1+x^2}}}$$

$$= \sqrt{\frac{1-\sqrt{1-x^2}}{1+\sqrt{1+x^2}}} \tag{4.3}$$

Thus, (4.3) is the radical form to the half of given expression (4.1).

2. Next, we need to derive the outcome of (4.2). First, by replacing x with sl(x), we can rewrite (4.3) as a form of the half angle:

$$\mathrm{sl}\left(\frac{x}{2}\right) = \sqrt{\frac{1-\sqrt{1-sl^2(x)}}{1+\sqrt{1+sl^2(x)}}} \tag{4.4}$$

3. Hence, continue replacing x with x/2 gives

$$\mathrm{sl}\left(\frac{x}{4}\right) = \sqrt{\frac{1-\sqrt{1-sl^2\left(\dfrac{x}{2}\right)}}{1+\sqrt{1+sl^2\left(\dfrac{x}{2}\right)}}} \tag{4.5}$$

4. Applying (4.4) in (4.5) and replacing x with arcsl(x) gives

$$\mathrm{sl}\left(\frac{1}{4}\,\mathrm{arcsl}(x)\right) = \sqrt{\frac{1-\sqrt{1-\dfrac{1-\sqrt{1-x^2}}{1+\sqrt{1+x^2}}}}{1+\sqrt{1+\dfrac{1-\sqrt{1-x^2}}{1+\sqrt{1+x^2}}}}} \tag{4.6}$$

Simplifying further gives the radical form of the quarter of the lemniscate sine:

$$\mathrm{sl}\left(\frac{1}{4}\,\mathrm{arcsl}(x)\right) = \sqrt{\frac{\sqrt{1+\sqrt{1+x^2}}-\sqrt{\sqrt{1+x^2}+\sqrt{1-x^2}}}{\sqrt{1+\sqrt{1+x^2}}+\sqrt{2+\sqrt{1+x^2}-\sqrt{1-x^2}}}} \; . \tag{4.7}$$

5. Repeat the step 2, replacing x with x/2 into (4.6) and applying (4.4), we get

$$\mathrm{sl}\left(\frac{x}{8}\right)=\sqrt{\frac{1-\sqrt{1-\dfrac{1-\sqrt{1-\dfrac{1-\sqrt{1-sl^2x}}{1+\sqrt{1+sl^2x}}}}{1+\sqrt{1+\dfrac{1-\sqrt{1-sl^2x}}{1+\sqrt{1+sl^2x}}}}}}{1+\sqrt{1+\dfrac{1-\sqrt{1-\dfrac{1-\sqrt{1-sl^2x}}{1+\sqrt{1+sl^2x}}}}{1+\sqrt{1+\dfrac{1-\sqrt{1-sl^2x}}{1+\sqrt{1+sl^2x}}}}}}}} \qquad (4.8)$$

Then substituting x by arcsl(x) gives

$$\mathrm{sl}\left(\frac{1}{8}\mathrm{arcsl(x)}\right)=\sqrt{\frac{1-\sqrt{1-\dfrac{1-\sqrt{1-\dfrac{1-\sqrt{1-x^2}}{1+\sqrt{1+x^2}}}}{1+\sqrt{1+\dfrac{1-\sqrt{1-x^2}}{1+\sqrt{1+x^2}}}}}}{1+\sqrt{1+\dfrac{1-\sqrt{1-\dfrac{1-\sqrt{1-x^2}}{1+\sqrt{1+x^2}}}}{1+\sqrt{1+\dfrac{1-\sqrt{1-x^2}}{1+\sqrt{1+x^2}}}}}}}} \qquad (4.9)$$

By simplifying further, we deduce that

$$\mathrm{sl}\left(\frac{1}{8}\mathrm{arcsl(x)}\right)=\sqrt{\frac{\sqrt{\sqrt{1+\sqrt{1+x^2}}+\sqrt{2}}-\sqrt{\sqrt{2\sqrt{1-x^2}}+\sqrt{2}}}{\sqrt{\sqrt{1+\sqrt{1+x^2}}+\sqrt{2}}+\sqrt{2\sqrt{1+\sqrt{1+x^2}}-\sqrt{2\sqrt{1-x^2}}+\sqrt{2}}}} \qquad (4.10)$$

Continuing replacing x with x/2 and scaling by a factor 1/2^n, we derive a set of nested radical identities for nth radical, as illustrated below:

$$\text{sl}\left(\frac{1}{2^n}\text{arcsl}(x)\right) = \sqrt{\frac{1-\sqrt{1-\dfrac{1-\sqrt{1-\dots\sqrt{1-\dfrac{1-\sqrt{1-x^2}}{1+\sqrt{1+x^2}}}}}{1+\sqrt{1+\dots\sqrt{1+\dfrac{1-\sqrt{1-x^2}}{1+\sqrt{1+x^2}}}}}}}{1+\sqrt{1+\dfrac{1-\sqrt{1-\dots\sqrt{1-\dfrac{1-\sqrt{1-x^2}}{1+\sqrt{1+x^2}}}}}{1+\sqrt{1+\dots\sqrt{1+\dfrac{1-\sqrt{1-x^2}}{1+\sqrt{1+x^2}}}}}}}} \tag{4.11}$$

(It consists of n nested square roots, excluding the square root of $(1\pm x^2)$.)

Noting that we also investigate for the half and quarter of the lemniscate cosine function and its inverse, denoted as cl and arccl. It yields comparable results, however, their algebraic patterns do not exhibit a certain symmetry as the sl function. The following outcomes are presented for reference.

$$\text{cl}\left(\frac{1}{2}\text{arccl}(x)\right) = \sqrt{\frac{\sqrt{1+x^2}+\sqrt{2}\,x}{\sqrt{1+x^2}+\sqrt{2}}} \tag{4.12}$$

$$\text{cl}\left(\frac{1}{4}\text{arccl}(x)\right) = \sqrt{\frac{2\sqrt{1+x^2}+\sqrt{2(1+x^2)}+\sqrt{2}\sqrt{\sqrt{1+x^2}+\sqrt{2}\,x}+2\sqrt{2}\,x+\sqrt{2}+2}{2\sqrt{1+x^2}+\sqrt{2(1+x^2)}+2\sqrt{2}\,x+\sqrt{2}+2}} \tag{4.13}$$

Proof. See Appendix Miscellaneous Proofs, [1-B-2].

Example. Evaluate integral $\displaystyle\int \cos\left(\frac{1}{4}\cos^{-1}(x)\right)dx$ $\tag{4.14}$

Solution. The integral (4.14) can be challenging to evaluate using trigonometric functions in Calculus. Readers may utilize a powerful computational tool such as

Wolfram or Matlab, as the result can be very complicated at the time of writing this book. To solve this integral, we use (1.5) in the following steps:

$$\int \cos\left(\frac{1}{4}\cos^{-1}(x)\right)dx = \int \frac{1}{2}\sqrt{2+\sqrt{2+2x}}\,dx \qquad (4.14\,a)$$

Let $u=\sqrt{2x+2}$. Then take derivative,

$$du = \frac{1}{\sqrt{2x+2}}dx \Rightarrow dx = u\,du.$$

By substituting in (4.12a), which gives

$$=\frac{1}{2}\int \sqrt{u+2}\,u\,du \qquad (4.14\,b)$$

Let $t=\sqrt{u+2} \Rightarrow u=t^2-2$. Then du = 2tdt.

Substituting u and du into (4.14b) yields the following.

$$=\frac{1}{2}\int t\left(t^2-2\right)2t\,dt$$

$$=\int \left(t^4-2t^2\right)dt$$

$$=\frac{t^5}{5}-\frac{2}{3}t^3+C$$

$$=\frac{1}{5}(u+2)^{5/2}-\frac{2}{3}(u+2)^{3/2}+C$$

$$=\frac{1}{5}\left(\sqrt{2(x+1)}+2\right)^{5/2}-\frac{2}{3}\left(\sqrt{2(x+1)}+2\right)^{3/2}+C$$

Thus, $\int \cos\left(\frac{1}{4}\cos^{-1}(x)\right)dx = \frac{1}{5}\left(\sqrt{2(x+1)}+2\right)^{5/2}-\frac{2}{3}\left(\sqrt{2(x+1)}+2\right)^{3/2}+C.$

Relationship Between Trigonometric Functions and Inverse Trigonometric Function Table

Each entry of the following Table II shows the composition formula for a specific trigonometric function and its inverse for positive real value of x with the *scaling angles* 1, ½ and ¼.

Table II – Relationship for Scaling Angles 1, ½ and ¼

$$\cos\left(\cos^{-1}(x)\right)=x$$

$$\cos\left(\frac{1}{2}\cos^{-1}(x)\right)=\sqrt{\frac{1+x}{2}}=\frac{1}{2}\sqrt{2+2x}$$

$$\cos\left(\frac{1}{4}\cos^{-1}(x)\right)=\frac{1}{2}\sqrt{2+\sqrt{2+2x}}$$

$$\sin\left(\sin^{-1}(x)\right)=x$$

$$\sin\left(\frac{1}{2}\sin^{-1}(x)\right)=\frac{1}{2}\sqrt{2-2\sqrt{1-x^2}}$$

$$\sin\left(\frac{1}{4}\sin^{-1}(x)\right)=\frac{1}{2}\sqrt{2-\sqrt{2+2\sqrt{1-x^2}}}$$

$$\sec\left(\cos^{-1}(x)\right)=\frac{1}{x}$$

$$\sec\left(\frac{1}{2}\cos^{-1}(x)\right)=\frac{\sqrt{2}}{\sqrt{1+x}}$$

$$\sec\left(\frac{1}{4}\cos^{-1}(x)\right)=\frac{2}{\sqrt{2+\sqrt{2+2}}}$$

$$\csc\left(\sin^{-1}(x)\right)=\frac{1}{x}$$

$$\csc\left(\frac{1}{2}\sin^{-1}(x)\right)=\frac{2}{\sqrt{2-2\sqrt{1-x^2}}}$$

$$\csc\left(\frac{1}{4}\sin^{-1}x\right)=\frac{2}{\sqrt{2-\sqrt{2+2\sqrt{1-x^2}}}}$$

$$\tan\left(\tan^{-1}(x)\right)=x$$

$$\tan\left(\frac{1}{2}\tan^{-1}(x)\right)=\sqrt{\frac{\sqrt{1+x^2}-1}{\sqrt{1+x^2}+1}}=\frac{\sqrt{1+x^2}-1}{x}$$

$$\cot\left(\tan^{-1}(x)\right)=\frac{1}{x}$$

$$\cot\left(\frac{1}{2}\tan^{-1}(x)\right)=\sqrt{\frac{\sqrt{1+x^2}+1}{\sqrt{1+x^2}-1}}=\frac{\sqrt{1+x^2}+1}{x}$$

$$\tan\left(\frac{1}{4}\tan^{-1}(x)\right)=\sqrt{\frac{2\sqrt[4]{1+x^2}-\sqrt{2+2\sqrt{1+x^2}}}{2\sqrt[4]{1+x^2}+\sqrt{2+2\sqrt{1+x^2}}}}$$

$$\cot\left(\frac{1}{4}\tan^{-1}(x)\right)=\frac{\sqrt[4]{4\left(x^2+1\right)}+\sqrt{1+\sqrt{x^2+1}}}{\sqrt{\sqrt{x^2+1}-1}}$$

$$\cos\left(\sin^{-1}(x)\right)=\sqrt{1-x^2}$$

$$\cos\left(\frac{1}{2}\sin^{-1}(x)\right)=\frac{1}{2}\sqrt{2+2\sqrt{1-x^2}}$$

$$\cos\left(\frac{1}{4}\sin^{-1}x\right)=\frac{1}{2}\sqrt{2+\sqrt{2+2\sqrt{1-x^2}}}$$

$$\sin\left(\cos^{-1}(x)\right)=\sqrt{1-x^2}$$

$$\sin\left(\frac{1}{2}\cos^{-1}(x)\right)=\frac{1}{2}\sqrt{2-2x}$$

$$\sin\left(\frac{1}{4}\cos^{-1}x\right)=\frac{1}{2}\sqrt{2-\sqrt{2+2x}}$$

$$\tan\left(\sin^{-1}(x)\right)=\frac{x}{\sqrt{1-x^2}}$$

$$\tan\left(\frac{1}{2}\sin^{-1}(x)\right)=\frac{1-\sqrt{1-x^2}}{x}$$

$$\tan\left(\frac{1}{4}\sin^{-1}(x)\right)=\frac{\sqrt{2-\sqrt{1+\sqrt{1-x^2}}}}{\sqrt{1-\sqrt{1-x^2}}}$$

$$\tan\left(\cos^{-1}(x)\right)=\frac{\sqrt{1-x^2}}{x}$$

$$\tan\left(\frac{1}{2}\cos^{-1}(x)\right)=\sqrt{\frac{1-x}{1+x}}$$

$$\tan\left(\frac{1}{4}\cos^{-1}(x)\right)=\frac{\sqrt{2}-\sqrt{1+x}}{\sqrt{1-x}}$$

$$\cot\left(\sin^{-1}(x)\right)=\frac{\sqrt{1-x^2}}{x}$$

$$\cot\left(\frac{1}{2}\sin^{-1}(x)\right)=\frac{1+\sqrt{1-x^2}}{x}$$

$$\cot\left(\frac{1}{4}\sin^{-1}(x)\right)=\frac{\sqrt{2+\sqrt{1+\sqrt{1-x^2}}}}{\sqrt{1-\sqrt{1-x^2}}}$$

$$\cot\left(\cos^{-1}(x)\right)=\frac{x}{\sqrt{1-x^2}}$$

$$\cot\left(\frac{1}{2}\cos^{-1}(x)\right)=\sqrt{\frac{1+x}{1-x}}$$

$$\cot\left(\frac{1}{4}\cos^{-1}(x)\right)=\frac{\sqrt{1-x}}{\sqrt{2}-\sqrt{1+x}}$$

$$\sec\left(\sec^{-1}(x)\right)=x$$

$$\sec\left(\frac{1}{2}\sec^{-1}(x)\right)=\sqrt{\frac{2x}{1+x}}$$

$$\sec\left(\frac{1}{4}\sec^{-1}(x)\right)=\sqrt{\frac{2\sqrt{2x}}{\sqrt{1+x}+\sqrt{2x}}}$$

$$\csc\left(\csc^{-1}(x)\right)=x$$

$$\csc\left(\frac{1}{2}\csc^{-1}(x)\right)=\frac{2}{\sqrt{2-2\sqrt{1-\frac{1}{x^2}}}}=\sqrt{2x\left(x+\sqrt{x^2-1}\right)}$$

$$\csc\left(\frac{1}{4}\csc^{-1}(x)\right)=\frac{2}{\sqrt{2-\sqrt{2\left(1+\sqrt{1-\frac{1}{x^2}}\right)}}}$$

$$\sin\left(\sec^{-1}(x)\right)=\frac{\sqrt{x^2-1}}{x}$$

$$\sin\left(\frac{1}{2}\sec^{-1}(x)\right)=\sqrt{\frac{x-1}{2x}}$$

$$\sin\left(\frac{1}{4}\sec^{-1}(x)\right)=\sqrt{\frac{1}{2}\left(1-\sqrt{\frac{x+1}{2x}}\right)}$$

$$\sin\left(\csc^{-1}(x)\right)=\frac{1}{x}$$

$$\sin\left(\frac{1}{2}\csc^{-1}(x)\right)=\frac{1}{2}\sqrt{2-2\sqrt{1-\frac{1}{x^2}}}$$

$$\sin\left(\frac{1}{4}\csc^{-1}(x)\right)=\frac{1}{2}\sqrt{2-\sqrt{2+2\sqrt{1-\frac{1}{x^2}}}}$$

$$\cos\left(\sec^{-1}(x)\right)=\frac{1}{x}$$

$$\cos\left(\frac{1}{2}\sec^{-1}(x)\right)=\sqrt{\frac{x+1}{2x}}$$

$$\cos\left(\frac{1}{4}\sec^{-1}(x)\right)=\sqrt{\frac{1}{2}\left(1+\sqrt{\frac{x+1}{2x}}\right)}$$

$$\cos\left(\csc^{-1}(x)\right)=\frac{\sqrt{x^2-1}}{x}$$

$$\cos\left(\frac{1}{2}\csc^{-1}(x)\right)=\frac{1}{2}\sqrt{2+2\sqrt{1-\frac{1}{x^2}}}$$

$$\cos\left(\frac{1}{4}\csc^{-1}(x)\right)=\frac{1}{2}\sqrt{2+\sqrt{2+2\sqrt{1-\frac{1}{x^2}}}}$$

$$\tan\left(\sec^{-1}(x)\right)=\sqrt{x^2-1}$$

$$\tan\left(\frac{1}{2}\sec^{-1}(x)\right)=\sqrt{\frac{x-1}{x+1}}$$

$$\tan\left(\frac{1}{4}\sec^{-1}(x)\right)=\frac{\sqrt{x-1}}{\sqrt{2x+\sqrt{x+1}}}$$

$$\tan\left(\csc^{-1}(x)\right)=\frac{1}{\sqrt{x^2-1}}$$

$$\tan\left(\frac{1}{2}\csc^{-1}(x)\right)=\frac{1}{x+\sqrt{x^2-1}}$$

$$\tan\left(\frac{1}{4}\csc^{-1}(x)\right)=\frac{\sqrt{x-\sqrt{x^2-1}}}{\sqrt{2x+\sqrt{x+\sqrt{x^2-1}}}}$$

$$\cot\left(\sec^{-1}(x)\right)=\frac{1}{\sqrt{x^2-1}}$$

$$\cot\left(\frac{1}{2}\sec^{-1}(x)\right)=\sqrt{\frac{x+1}{x-1}}$$

$$\cot\left(\frac{1}{4}\sec^{-1}(x)\right)=\frac{\sqrt{2x+\sqrt{x+1}}}{\sqrt{x-1}}$$

$$\cot\left(csc^{-1}(x)\right)=\sqrt{x^2-1}$$

$$\cot\left(\frac{1}{2}\csc^{-1}(x)\right)=x+\sqrt{x^2-1}$$

$$\cot\left(\frac{1}{4}\csc^{-1}(x)\right)=\frac{\sqrt{2x+\sqrt{x+\sqrt{x^2-1}}}}{\sqrt{x-\sqrt{x^2-1}}}$$

Appendix 1-A: Various Proofs Relate Trigonometric Functions and Inverse Trigonometric Functions

We provide various proofs for selected formulas that relate trigonometric functions and inverse trigonometric functions in Table II. These formulas are arranged in sequential order, where the first formula serves as the basis for proving the subsequent formulas in the order presented.

1. $\cos\left(\cos^{-1}(x)\right)=x$ $\hspace{3cm}$ (1)

2. $\sin\left(\cos^{-1}(x)\right)=\sqrt{1-x^2}$ $\hspace{3cm}$ (2)

By setting y = cos⁻¹(x), we can deduce that x = cos(y). Upon squaring both sides, we obtain

$$x^2=\cos^2(y).$$

Applying Pythagorean trigonometric identity, we have

$$\sin^2(y)=1-x^2$$

Taking square root on both sides and after that substituting y back in, we obtain the result,

$$\sin\left(\cos^{-1}(x)\right)=\sqrt{1-x^2}.$$

3. $\cos\left(\sin^{-1}(x)\right)=\sqrt{1-x^2}$ $\hspace{3cm}$ (3)

Take inverse sine of both sides of (2), we have

$$\cos^{-1}(x)=\sin^{-1}\left(\sqrt{1-x^2}\right).$$

Replacing x with $\sqrt{1-x^2}$ gives

$$\cos^{-1}\left(\sqrt{1-x^2}\right)=\sin^{-1}(x).$$

Then take cosine both sides, which gives

$$\cos\left(\sin^{-1}(x)\right)=\sqrt{1-x^2}.$$

4. $$\tan\left(\sin^{-1}(x)\right)=\frac{x}{\sqrt{1-x^2}} \qquad (4)$$

By using the definition of tangent function, we have

$$\tan\left(\sin^{-1}(x)\right)=\frac{\sin\left(\sin^{-1}(x)\right)}{\cos\left(\sin^{-1}(x)\right)}=\frac{x}{\sqrt{1-x^2}}$$

5. $$\sin\left(\tan^{-1}(x)\right)=\frac{x}{\sqrt{1+x^2}} \qquad (5)$$

From (4), we can get

$$\sin\left(\tan^{-1}\frac{x}{\sqrt{1-x^2}}\right)=x \qquad (5a)$$

By setting $u=\frac{x}{\sqrt{1-x^2}}$,

we deduce that $u^2(1-x^2) = x^2$ then rewrite as $x^2(1+u^2) = u^2$. Take square root of both sides gives

$$x=\frac{u}{\sqrt{1+u^2}}.$$

Substitute x into (5a) gives $\sin\left(\tan^{-1}(u)\right)=\frac{u}{\sqrt{1+u^2}}$. Rename u as x, and it follows the result in (5).

6. $$\sin\left(\tan^{-1}(x)\right)=\frac{1}{\sqrt{1+x^2}} \qquad (6)$$

By squaring both sides of (5) and applying the Pythagorean trigonometric identity, which leads to the desired result.

7. $\tan\left(\tan^{-1}(x)\right)=x$ (7)

8. $\cot\left(\tan^{-1}(x)\right)=\dfrac{1}{x}$ (8)

9. $\cot\left(\sin^{-1}(x)\right)=\dfrac{\sqrt{1-x^2}}{x}$ (9)

By applying the identity cot(x) = 1/tan(x) in (4), which follows the desired result.

10. $\sin\left(\cot^{-1}(x)\right)=\dfrac{1}{\sqrt{1+x^2}}$ (10)

By following a procedure similar to the one described in (5), we can achieve the desired outcome.

11. $\cos\left(\cot^{-1}(x)\right)=\dfrac{x}{\sqrt{1+x^2}}$ (11)

By squaring both sides of (10) and applying the Pythagorean trigonometric identity, we obtain the result (11).

12. $\sec\left(\sec^{-1}(x)\right)=x$ (12)

13. $\cos\left(\sec^{-1}(x)\right)=\dfrac{1}{x}$ (13)

14. $\sin\left(\sec^{-1}(x)\right)=\dfrac{\sqrt{x^2-1}}{x}$ (14)

By squaring both sides of (13) and applying the Pythagorean trigonometric identity, we have

$$\sin^2\left(\sec^{-1}(x)\right)=1-\frac{1}{x^2}.$$

Take square root of both sides gives

$$\sin\left(\sec^{-1}(x)\right) = \frac{\sqrt{x^2-1}}{x}.$$

15. $\quad \tan\left(\sec^{-1}(x)\right) = \sqrt{x^2-1}$ \hfill (15)

By using (13) and (14), we derive

$$\tan\left(\sec^{-1}(x)\right) = \frac{\sin\left(\sec^{-1}(x)\right)}{\cos\left(\sec^{-1}(x)\right)} = \sqrt{x^2-1}.$$

16. $\quad \sec\left(\cos^{-1}(x)\right) = \frac{1}{x}$ \hfill (16)

From (12), we have

$$\sec^{-1}(x) = \cos^{-1}\left(\frac{1}{x}\right)$$

$$\sec\left(\cos^{-1}\left(\frac{1}{x}\right)\right) = x \qquad \text{(Replace x with 1/x.)}$$

$$\sec\left(\cos^{-1}(x)\right) = \frac{1}{x}$$

17. $\quad \csc\left(\sec^{-1}(x)\right) = \frac{x}{\sqrt{x^2-1}}$ \hfill (17)

The result follows from formula (14).

18. $\quad \csc\left(\csc^{-1}(x)\right) = x$ \hfill (18)

19. $\quad \sin\left(\csc^{-1}(x)\right) = \frac{1}{x}$ \hfill (19)

20. $\quad \cos\left(\csc^{-1}(x)\right) = \frac{\sqrt{x^2-1}}{x}$ \hfill (20)

Square both sides of (19) and apply Pythagorean trigonometric identity gives

$$\cos^2\left(\csc^{-1}(x)\right) = \frac{x^2-1}{x^2}$$

$$\cos\left(csc^{-1}(x)\right)=\frac{\sqrt{x^2-1}}{x}$$

21. $\tan\left(csc^{-1}(x)\right)=\dfrac{1}{\sqrt{x^2-1}}$ \hfill (21)

The result is derived from (19) and (20).

22. $\cot\left(csc^{-1}(x)\right)=\sqrt{x^2-1}$ \hfill (22)

The result is derived by using (21) and applying the identity tan(x) = 1/cot(x).

Appendix 1-B: Miscellaneous Symbols and Proofs

[1-B-1]

The following symbol notations are used in this book: $\mathbb{N}$ for natural numbers, $\mathbb{Z}$ for integers, $\mathbb{Q}$ for rationals, $\mathbb{R}$ for real numbers, and $\mathbb{C}$ for complex numbers.

[1-B-2]

To prove (4.12), we use two relations that involve lemniscate sine sl and lemniscate cosine cl functions [6] as follows:

$$sl^2 x = \frac{1 - cl^2 x}{1 + cl^2 x}$$

$$cl^2 \frac{x}{2} = \frac{cl\, x \sqrt{1 + sl^2 x} + 1}{\sqrt{1 + sl^2 x} + 1}$$

Substituting x by arccl(x) to both formulas above gives

$$sl^2\left(arccl\left(x\right)\right) = \frac{1 - x^2}{1 + x^2} \tag{1-C-1}$$

$$cl^2 \frac{x}{2} = \frac{x\sqrt{1 + sl^2\left(arccl\left(x\right)\right)} + 1}{\sqrt{1 + sl^2\left(arccl\left(x\right)\right)} + 1} \tag{1-C-2}$$

Substituting expression (1-C-1) in (1-C-2) gives

$$cl^2\left(\frac{1}{2} arccl\left(x\right)\right) = \frac{x\sqrt{1 + \dfrac{1 - x^2}{1 + x^2}} + 1}{\sqrt{1 + \dfrac{1 - x^2}{1 + x^2}} + 1} \tag{1-C-3}$$

Then take square root of both sides after further deduction leads to result (4.12). Similarly, we can obtain (4.13) by replacing x with arccl(x) into (4.12) and deducing expressions on both nominator and denominator.

Chapter 2

Innovative Approaches to Explore Polynomial Equation Solving: Trigonometric and Hyperbolic Insights

This chapter presents a consistent approach to solving general quadratic, cubic, quartic, and other special higher order equations. The approach involves using compositions of trigonometric functions and inverse trigonometric functions, as well as hyperbolic trigonometric functions and their inverse functions. The chapter is divided into four sections. The first three sections deal with solutions to general quadratic, cubic, and quartic equations. The fourth section offers solutions to other higher degree equations in some special forms. Each section includes a derivation of the solution, examples, and a summary for the corresponding type of polynomial equation.

Section I. Solving General Quadratic Equation

We can solve a general quadratic equation by leveraging the composition of the double angle cosine with its inverse. This composition yields a second degree equation. Our technique involves transforming this second degree equation into the general form of a quadratic equation that matches the desired coefficients. Simultaneously, we show how to find solutions during this transformation process.

The general form of a quadratic equation [7] is given by:

$$a x^2 + b x + c = 0 \quad (a \neq 0),$$

(1)

where a, b, and c are real numbers.

A. Solution to Quadratic Equation (Type 1)

To solve equation (1), we start by using the double angle identity for cosine, which can be expressed as $\cos 2x = 2\cos^2 x - 1.$ By substituting x with $\cos^{-1}(x)$, we obtain a composite form of the double cosine with its inverse, which results in

$$\cos\left(2\cos^{-1}(x)\right) = 2x^2 - 1, \quad -1 \leqslant x \leqslant 1. \tag{2}$$

[for other trigonometric identities, see Appendix 2-A-1.]

The expression on the right-side of equation (2) can be classified as a second degree equation due to the highest power of the variable x being 2. Our subsequent objective is to transform the right-hand side of equation (2) back into equation (1) and determine its solution by solving the trigonometric equation. The step-by-step procedure is outlined as follows:

- We want to have $a^2 x$ and bx terms appearing on the right-hand side of equation (2) by substituting x with $\sqrt{a/2}x + b$ into both sides of (2), which gives

$$\cos\left(2\cos^{-1}\left(\sqrt{\frac{a}{2}}x + b\right)\right) = 2\left(\sqrt{\frac{a}{2}}x + b\right)^2 - 1$$

$$= ax^2 + 2\sqrt{2a}\,bx + 2b^2 - 1. \tag{3}$$

- Next, we remove $2\sqrt{2a}$ from x term by replacing b with $b/(2\sqrt{2a})$ in both sides of (3), which gives

$$\cos\left(2\cos^{-1}\left(\sqrt{\frac{a}{2}}x + \frac{b}{2\sqrt{2a}}\right)\right) = ax^2 + bx + \frac{b^2}{4a} - 1. \tag{4}$$

- In order to have c term appearing on the right-hand side of (4), we introduce m as a real number. By subtracting m from both sides of (4), which gives

$$\cos\left(2\cos^{-1}\left(\sqrt{\frac{a}{2}}x + \frac{b}{2\sqrt{2a}}\right)\right) - m = ax^2 + bx + \frac{b^2}{4a} - 1 - m. \tag{5}$$

- Therefore, we can establish a system of two simultaneous equations by setting both sides of (5) to zero:

$$\begin{cases} a x^2 + b x + \dfrac{b^2}{4a} - 1 - m = 0 & (6a) \\[2em] \cos\left(2\cos^{-1}\left(\sqrt{\dfrac{a}{2}}\, x + \dfrac{b}{2\sqrt{2a}}\right)\right) - m = 0. & (6b) \end{cases} \quad (6)$$

- Now let

$$\frac{b^2}{4a} - 1 - m = c,$$

then solving for m, we have

$$m = \frac{b^2}{4a} - c - 1.$$

- Substituting m into (6a) and (6b) gives

$$\begin{cases} a x^2 + b x + c = 0 & (7a) \\[2em] \cos\left(2\cos^{-1}\left(\sqrt{\dfrac{a}{2}}\, x + \dfrac{b}{2\sqrt{2a}}\right)\right) - \dfrac{b^2}{4a} + c + 1 = 0. & (7b) \end{cases} \quad (7)$$

We have established a system of two simultaneous equations (7) in which equation (7a) is the general form of the quadratic equation transformed from equation (2), and (7b) is the trigonometric equation with the variable x that needs to be solved. It is worth noting that x is not only the roots of (7b), but also the roots of (7a), since both equations are equivalent.

Next, we solve for x from trigonometric equation (7b) as follows:
- Isolate the cosine function by moving other terms to the right-hand side of the equal sign, we have

$$\cos\left(2\cos^{-1}\left(\sqrt{\frac{a}{2}}\, x + \frac{b}{2\sqrt{2a}}\right)\right) = \frac{b^2}{4a} - c - 1.$$

- Take the inverse cosine of both sides:

$$2\cos^{-1}\left(\sqrt{\frac{a}{2}}\, x + \frac{b}{2\sqrt{2a}}\right) = \cos^{-1}\left(\frac{b^2}{4a} - c - 1\right).$$

- Divide both sides by 2, and then take the cosine of both sides, which gives

$$\sqrt{\frac{a}{2}}\,x+\frac{b}{2\sqrt{2}\,a}=\cos\left(\frac{1}{2}\cos^{-1}\left(\frac{b^2}{4a}-c-1\right)\right).$$

- Isolate $\sqrt{a/2}\,x$ by subtracting $b/(2\sqrt{2}a)$ from both sides. Then solve for x by dividing through by $\sqrt{a/2}$, which gives the solution

$$x=\sqrt{\frac{2}{a}}\cos\left(\frac{1}{2}\cos^{-1}\left(\frac{b^2}{4a}-c-1\right)\right)-\frac{b}{2a}.$$

- Since the cosine function is periodic with a period of 2π, we can express the solution for x in terms of its periodicity [see Appendix 2-A-2] as

$$x_{k+1}=\sqrt{\frac{2}{a}}\cos\left(\frac{1}{2}\cos^{-1}\left(\frac{b^2}{4a}-c-1\right)+k\pi\right)-\frac{b}{2a},\quad k\in\mathbb{Z}=\{0,\pm1,\pm2,\dots\}.$$

Therefore, rewriting (7) gives the general solution of the quadratic equation, which is presented

$$\begin{cases}a x^2+b x+c=0\ (a\neq0) & (8a)\\ x_{k+1}=\sqrt{\dfrac{2}{a}}\cos\left(\dfrac{1}{2}\cos^{-1}\left(\dfrac{b^2}{4a}-c-1\right)+k\pi\right)-\dfrac{b}{2a},\ k\in\mathbb{Z}=\{0,\pm1,\pm2,\dots\}. & (8b)\end{cases}\quad(8)$$

$$\text{for }\left|\frac{b^2}{4a}-c-1\right|\leq1.$$

Please note that going forward, we will adopt this format to present the equation to be solved alongside its solution as a convenient means of conveying explanations to our kindle readers.

Using equation (8b) with k = 0 and k = 1, we obtain two roots:

$$\begin{cases}a x^2+b x+c=0 & (9a)\\ x_{1,2}=\pm\sqrt{\dfrac{2}{a}}\cos\left(\dfrac{1}{2}\cos^{-1}\left(\dfrac{b^2}{4a}-c-1\right)\right)-\dfrac{b}{2a}\ \text{for}\left|\dfrac{b^2}{4a}-c-1\right|\leq1. & (9b)\end{cases}\quad(9)$$

Thus, we have demonstrated how the equation of the right-hand side of (2) can be transformed into the general form of quadratic equation (9a) and how its solution can be derived as indicated in (9b). However, the solution (9b) is restricted to the domain of cos⁻¹(x) in the closed interval [-1, 1]. To find solution to other domain (-∞, -1] or [1, ∞), we can use the double angle identity for hyperbolic cosine and its inverse [see Appendix 2-A-3]:

$$\cosh\left(2\cosh^{-1}(x)\right)=2\,x^2-1. \tag{10}$$

Since the expressions on the right-hand side of formulas (2) and (10) are identical, by using a similar procedure as starting from formula (2) to apply on formula (10), we expect solution is analogous to (9b) but it contains hyperbolic functions instead. As a result, the solution of the quadratic equation is derived and presented as follows:

$$\begin{cases} a\,x^2+b\,x+c=0 & (11a) \\ x_{k+1}=\sqrt{\dfrac{2}{a}}\cosh\left(\dfrac{1}{2}\cosh^{-1}\left(\dfrac{b^2}{4\,a}-c-1\right)+k\,\pi i\right)-\dfrac{b}{2\,a} & (11b) \end{cases} \tag{11}$$

$$\text{for}\quad \frac{b^2}{4\,a}-c-1\geq 1 \Leftrightarrow \frac{b^2}{4\,a}-c\geq 2,\quad k\in\mathbb{Z}, \text{ and i is the imaginary unit.}$$

In the upcoming sections, we will use the following identities to help in mathematical manipulations while solving equations:

Table 1 - Relations of Trigonometric and Hyperbolic Trigonometric Functions

$\cosh(x+2\pi i)=\cosh(x)$	(A)	$\sinh^{-1}(ix)=i\sin^{-1}(x)$	(G)
$\cosh^{-1}(-x)=\pi i-\cosh^{-1}(x)$	(B)	$\cosh(ix)=\cos(x)$	(H)
$\cosh(x\pm\pi i/2)=\pm i\sinh(x)$	(C)	$\cos(ix)=\cosh(x)$	(I)
$\cosh(x\pm\pi i)=-\cosh(x)$	(D)	$\tan^{-1}i\,x=i\tanh^{-1}x$	(K)
$\sinh(x\pm\pi i)=-\sinh(x)$	(E)	$\tan(ix)=i\tanh(x)$	(L)
$\sinh(ix)=i\sin(x)$	(F)		

Although we work with coefficients assumed to be real numbers, the introduction of complex coefficients, often due to even radicals, is possible. The relations in Table 1 involve the utilization of the complex unit i to transform trigonometric functions into their hyperbolic counterparts or vice versa [see Appendix 2-A-0.]. Our focus in this chapter remains on solving equations within the realm of real numbers, excluding equations with coefficients from the complex field, which fall beyond the scope of this book.

Using (11b) with k = 0 and k = 1, and applying formula (D), respectively, we explicitly obtain two roots:

$$\begin{cases} a\,x^2+b\,x+c=0 & (12a) \\ x_{1,2}=\pm\sqrt{\dfrac{2}{a}}\cosh\left(\dfrac{1}{2}\cosh^{-1}\left(\dfrac{b^2}{4\,a}-c-1\right)\right)-\dfrac{b}{2\,a} & (12b) \end{cases} \tag{12}$$

for $\dfrac{b^2}{4a} - c - 1 \geq 1$.

Since the inverse hyperbolic cosine has domain $[1, \infty)$, we can achieve a mirror of this domain to $(-\infty, 1]$ using the identity $\cosh^{-1}(-x) = i\pi - \cosh^{-1}(x)$ when $x \leq -1$. We will explicitly show formula for this case when we examine some cases of expression $(b^2/(4a) - c - 1)$ in (12b).

Special Cases of $(b^2/(4a) - c - 1)$ in (12b)

➤ If $\dfrac{b^2}{4a} - c - 1 \geq 1 \Leftrightarrow \dfrac{b^2}{4a} - c \geq 2$, formula (12b) provides two real roots.

➤ If $\dfrac{b^2}{4a} - c - 1 \leq -1 \Leftrightarrow \dfrac{b^2}{4a} - c \leq 0$, formula (12b) provides two complex conjugate roots. We can prove it by applying formula (B) to (12b) as follows:

$$x_{1,2} = \pm\sqrt{\frac{2}{a}}\cosh\left(\frac{1}{2}\cosh^{-1}\left(\frac{b^2}{4a} - c - 1\right)\right) - \frac{b}{2a}$$

$$= \pm\sqrt{\frac{2}{a}}\cosh\left(\frac{i\pi}{2} - \frac{1}{2}\cosh^{-1}\left(-\left[\frac{b^2}{4a} - c - 1\right]\right)\right) - \frac{b}{2a} \qquad (13)$$

By applying formula (C), we deduce that

$$= \pm i\sqrt{\frac{2}{a}}\sinh\left(\frac{1}{2}\cosh^{-1}\left(-\frac{b^2}{4a} + c + 1\right)\right) - \frac{b}{2a}, \qquad (14)$$

which are two complex conjugate roots of equation (12a).

We have demonstrated how to construct quadratic equation (1) from the composition of the cosine function with its inverse, or from the composition of the hyperbolic cosine with its inverse, and how to find its solutions (9b), (12b) and (14b) that include real and complex roots.

Summary of Solution to General Quadratic Equation (Type 1)

Quadratic equation: $ax^2 + bx + c = 0 \ (a \neq 0)$

Define $T \equiv \dfrac{b^2}{4a} - c - 1$.

The solution to the general quadratic equation can be expressed as follows:

+> If $|T| \leq 1$, the equation has two real roots,

$$x_{1,2} = \pm\sqrt{\frac{2}{a}}\cos\left(\frac{1}{2}\cos^{-1}(T)\right) - \frac{b}{2a}.$$

+> If $T \geq 1$, the equation has two real roots,

$$x_{1,2} = \pm\sqrt{\frac{2}{a}}\cosh\left(\frac{1}{2}\cosh^{-1}(T)\right) - \frac{b}{2a}.$$

+> If $T \leq -1$, the equation has two complex conjugate roots

$$x_{1,2} = \pm i\sqrt{\frac{2}{a}}\sinh\left(\frac{1}{2}\cosh^{-1}(-T)\right) - \frac{b}{2a}.$$

Note that if $T = 1$ or $T = -1$, all above formulas give repeated roots, namely

$$x_1 = x_2 = -\frac{b}{2a}.$$

Notes:

1. The reason we derive formula (14b) is because the domain of inverse hyperbolic cosine is defined only within the interval $[1, \infty)$. The major of the current calculators, however, are constrained to operate to other interval $(-\infty, -1]$, which is supposed to give complex results. In fact, the formulas (8) and (11) are suffice to represent the solution of the general quadratic equation (1) once the calculators are equipped to compute values of the inverse hyperbolic functions for other complementary intervals. Notice that when working with complex coefficients, there's no need to impose any domain conditions in those solution formulas.

2. Although the use of trigonometry to solve quadratic equations may seem more complicated than the traditional method at first, it has its own benefits. In later sections, when we deal with higher degree equations, once we have found one root, the remaining roots will follow naturally due to the periodicity of trigonometric functions.

3. Formula (9) or (11) represents the solution of equation (1), and it can be expressed in radicals by applying formula (1.3) or (1.2.2a) found in Chapter 1.

B. Solution to General Quadratic Equation (Type 2)

In this section, we use another trigonometric identity [see Appendix 2-A-4],

$$\cos\left(4\sin^{-1}(x)\right)=8x^4-8x^2+1,\tag{15}$$

to derive another solution formula for the quadratic equation by applying a similar process as previously described in part A. The following steps illustrate the derivation of this solution:

1. Let a, b, c and m be any real numbers. By replacing x with $\sqrt{x}/\sqrt{b}$ into (15), and multiplying through both its sides by $b^2/8$, which gives

$$\frac{b^2}{8}\cos\left(4\sin^{-1}\frac{\sqrt{x}}{\sqrt{b}}\right)=x^2-bx+\frac{b^2}{8}.\tag{16}$$

2. Then subtracting m from both sides of (16) gives

$$\frac{b^2}{8}\cos\left(4\sin^{-1}\frac{\sqrt{x}}{\sqrt{b}}\right)-m=x^2-bx+\frac{b^2}{8}-m\tag{17}$$

3. We establish a system of two simultaneous equations by setting both sides of (17) to zero:

$$\begin{cases}x^2-bx+\dfrac{b^2}{8}-m=0 & (18\,a)\\[2em]\dfrac{b^2}{8}\cos\left(4\sin^{-1}\left(\dfrac{\sqrt{x}}{\sqrt{b}}\right)\right)-m=0 & (18\,b)\end{cases}\tag{18}$$

4. Solve for x from trigonometric equation (18b) as follows:
   ➢ Rewrite equation (18b) as

$$\frac{b^2}{8}\cos\left(4\sin^{-1}\frac{\sqrt{x}}{\sqrt{b}}\right)=m.$$

   ➢ Then multiply both sides by $8/b^2$, which gives

$$\cos\left(4\sin^{-1}\frac{\sqrt{x}}{\sqrt{b}}\right)=\frac{8m}{b^2}.$$

➢ After applying the cosine function to both sides, proceed by dividing both sides by 4, resulting in.

$$\sin^{-1}\frac{\sqrt{x}}{\sqrt{b}}=\frac{1}{4}\cos^{-1}\left(\frac{8\,m}{b^2}\right).$$

➢ By taking sine both sides and then multiplying both sides by square root of b, which gives

$$\sqrt{x}=\sqrt{b}\sin\left(\frac{1}{4}\cos^{-1}\left(\frac{8\,m}{b^2}\right)\right).$$

➢ Taking square of both sides gives

$$x=b\sin^2\left(\frac{1}{4}\cos^{-1}\left(\frac{8\,m}{b^2}\right)\right).$$

➢ Applying double cosine angle gives

$$x=\frac{b}{2}\left[1-\cos\left(\frac{1}{2}\cos^{-1}\left(\frac{8\,m}{b^2}\right)\right)\right].$$

➢ Since the cosine function is periodic with a period of 2п, we rewrite the formula above in terms of periodicity,

$$x_{k+1}=\frac{b}{2}\left[1-\cos\left(\frac{1}{2}\cos^{-1}\left(\frac{8\,m}{b^2}\right)+k\,\pi\right)\right]\ ,\ \ k\in\mathbb{Z}.$$

While k belongs to the set of integers (k ∈ ℤ), it is sufficient to consider the values of k = 0 and k = 1.

5. As a result, equations (18) can be restated as

$$\begin{cases} x_{k+1}=\dfrac{b}{2}\left(1-\cos\left(\dfrac{1}{2}\cos^{-1}\left(\dfrac{8\,m}{b^2}\right)+k\,\pi\right)\right),k=0,1. & (19a)\\[4mm] x^2-b\,x+\dfrac{b^2}{8}-m=0. & (19b) \end{cases} \qquad (19)$$

6. We want c/a term appearing in (19b) by replacing (b²/8 – m) with c/a or m = (b²/8 – c/a) into (19a) and (19b), which gives

$$\begin{cases} x^2 - b\,x + \dfrac{c}{a} = 0 & (a \neq 0) \qquad (19c) \\[4mm] x_{k+1} = \dfrac{b}{2}\left(1 - \cos\left(\dfrac{1}{2}\cos^{-1}\left(1 - \dfrac{8c}{a\,b^2}\right) + k\,\pi\right)\right) & (19d) \end{cases}$$

7. We aim to introduce the term ax^2 into equation (19c). To achieve this, we substitute b with -b/a in equations (19c) and (19d). Next, we multiply both sides of equation (19c) by 'a' and further simplify. This process leads us to the roots of the general quadratic equation.

$$\begin{cases} a\,x^2 + b\,x + c = 0 & (a \neq 0) \qquad\qquad (20a) \\[4mm] x_{k+1} = -\dfrac{b}{2a}\left(1 - \cos\left[\dfrac{1}{2}\cos^{-1}\left(\dfrac{b^2 - 8ac}{b^2}\right) + \pi k\right]\right), & k = 0, 1, \qquad (20b) \end{cases} \qquad (20)$$

$$\text{for} \quad \left|\dfrac{b^2 - 8ac}{b^2}\right| \leq 1.$$

In a special case,

$$\cos\left[\dfrac{1}{2}\cos^{-1}\left(\dfrac{b^2 - 8ac}{b^2}\right)\right] = 0$$

$$\Rightarrow \cos^{-1}\left(\dfrac{b^2 - 8ac}{b^2}\right) = \pi$$

$$\Rightarrow b = \pm 2\sqrt{ac}.$$

Thus, when $b = \pm 2\sqrt{ac}$, the equation has a double root,

$$x_{1,2} = -\dfrac{b}{2a}.$$

Formula (20b) does not cover the solution of quadratic equation (20a) if $\left|\dfrac{b^2 - 8ac}{b^2}\right| \geq 1$. You might intuitively anticipate the formula for finding the roots of (20a) when $\dfrac{b^2 - 8ac}{b^2} \geq 1$ or when $\dfrac{b^2 - 8ac}{b^2} \leq -1$. It takes on a similar structure to formula (20b), but with hyperbolic functions. Indeed, by replacing x with ix into (15) and applying the identities sin(ix) = isinh(x) and cos(ix) = cosh(x), we deduce that

$$\cosh\left(4\sinh^{-1}(x)\right) = 8x^4 + 8x^2 + 1. \qquad (21)$$

The left-hand side of (21) is the composition of the quadruple angle for hyperbolic

76

cosine with inverse hyperbolic sine function. We observe that the results (15) and (21) are identical except the sign of $8x^2$ term. By following the same procedure as previously described, we can transform (21) into a system of two simultaneous equations in which the quadratic equation and its solution are resulted in,

$$\begin{cases} a x^2+b x+c=0 \quad (a\neq 0) & (22a) \\ x_{k+1}=-\dfrac{b}{2a}\left(1+\cosh\left[\dfrac{1}{2}\cosh^{-1}\left(\dfrac{b^2-8ac}{b^2}\right)+i\pi k\right]\right) \ for \ k=0,1. & (22b) \end{cases}$$ (22)

Using (22b) with k = 0 and k = 1, we explicitly obtain two roots:

$$\begin{cases} a x^2+b x+c=0 \quad (a\neq 0) & (23a) \\ x_{1,2}=-\dfrac{b}{2a}\left(1\pm\cosh\left[\dfrac{1}{2}\cosh^{-1}\left(\dfrac{b^2-8ac}{b^2}\right)\right]\right). & (23b) \end{cases}$$ (23)

[For other forms of (20b) and (23b), see Appendix 2-A-5]

Special Cases of (b²-8ac)/b² in (23b)

- If $\dfrac{b^2-8ac}{b^2}\geq 1 \Leftrightarrow ac\leq 0$, then equation (23a) has two real roots.

- If $\dfrac{b^2-8ac}{b^2}\leq -1 \Leftrightarrow b^2-4ac \leq 0$, then equation (23a) has two complex conjugate roots, namely

$$x_{1,2}=-\frac{b}{2a}\left(1\pm i\sinh\left[\frac{1}{2}\cosh^{-1}\left(\frac{8ac-b^2}{b^2}\right)\right]\right).$$ (24)

Summary of Solution to General Quadratic Equation (Type 2)

Quadratic equation: $a x^2+b x+c=0 \ (a\neq 0)$

Define $T=\dfrac{b^2-8ac}{b^2}$.

The solution to the general quadratic equation can be expressed as follows:

- If $|T|\leq 1$, the equation has two real roots, which are given by

$$x_{1,2}=-\frac{b}{2a}\left(1\pm\cos\left(\frac{1}{2}\cos^{-1}(T)\right)\right).$$

- If T≥1, the equation has two real roots, which are given by

$$x_{1,2} = -\frac{b}{2a}\left(1 \pm \cosh\left(\frac{1}{2}\cosh^{-1}(T)\right)\right).$$

- If T≤-1, the equation has two complex conjugate roots, which are given by

$$x_{1,2} = -\frac{b}{2a}\left(1 \pm i\sinh\left[\frac{1}{2}\cosh^{-1}(-T)\right]\right).$$

Note that T = -1, the equation has a repeated root, namely,

$$x_1 = x_2 = -\frac{b}{2a}.$$

We have encountered various derived solution formulas for the quadratic equation throughout our exploration, all of which are based on specific trigonometric or hyperbolic identities. There are still more solution formulas for the quadratic equation, as various forms of trigonometric functions or hyperbolic trigonometric identities exist. We will provide one of them, which is of tangent form, in the solving quartic equation section.

Section II. Solving General Cubic Equation

In this section, we use the same approach from Section I to solve for a general cubic equation. For simplicity, we refrain from presenting detailed step-by-step solutions for every mathematical expression or equation as we have done in Section I. Instead, we emphasize key results to simplify mathematical solutions, making it easier for readers to grasp and comprehend.

A. Solution to Cubic Equation (Type 1)

The general form of a cubic equation [8] is expressed as

$$a x^3 + b x^2 + c x + d = 0 \quad (a \neq 0), \tag{30}$$

where a, b, c and d are real numbers.

We use the triple angle identity for cosine to solve cubic equation (30). The triple angle identity for cosine is expressed as

$$\cos\left(3\cos^{-1}(x)\right) = 4x^3 - 3x. \tag{31}$$

Our goal now is to align the coefficients and terms in the equation on the right-hand side of (31) with those in equation (30). We aim to modify the coefficients in the equation on the right-hand side of (31) to mirror the structure and terms present in equation (30). and find its solution. Any changes made to the coefficients in the equation on the right-hand side of (31) will correspondingly impact its left-hand side. Consequently, the solution derived from the left-hand side of (31) will also serve as solution for equation (30). We demonstrate the process as shown in the following steps:

1. We want cx term appearing on the right-hand side of (31) by replacing x with x/(2c) into (31) and then multiplying through both its sides by $2c^3$, which gives

$$2c^3 \cos\left(3\cos^{-1}\frac{x}{2c}\right) = x^3 - 3c^2 x. \tag{32}$$

2. Continue replacing c with ($\sqrt{c}$)/($\sqrt{3}$) gives

$$\frac{2c\sqrt{c}}{3\sqrt{3}} \cos\left(3\cos^{-1}\frac{\sqrt{3}x}{2\sqrt{c}}\right) = x^3 - cx. \tag{33}$$

3. Assuming m be a real number. Then subtracting m from both sides of (33) gives

$$\frac{2c\sqrt{c}}{3\sqrt{3}} \cos\left(3\cos^{-1}\frac{\sqrt{3}x}{2\sqrt{c}}\right) - m = x^3 - cx - m. \tag{34}$$

4. Therefore, we can establish a system of two simultaneous equations by setting both sides of (34) to zero:

$$\begin{cases} x^3 - cx - m = 0 & (35\,a) \\ \dfrac{2c\sqrt{c}}{3\sqrt{3}} \cos\left(3\cos^{-1}\dfrac{\sqrt{3}x}{2\sqrt{c}}\right) - m = 0. & (35\,b) \end{cases} \tag{35}$$

5. Solving for x from (35b) gives the solution of cubic equation (35a)

$$x_{k+1} = \frac{2\sqrt{c}}{\sqrt{3}} \cos\left(\frac{1}{3}\cos^{-1}\left(\frac{3\sqrt{3}m}{2c\sqrt{c}}\right) + \frac{2\pi k}{3}\right), \quad k = 0,1,2. \tag{36}$$

6. We now want to have ax³ term appearing on (35a) by replacing m with m/a and c with c/a into (35a) and (36), and then multiplying both sides of (35a) by a, which gives

$$\begin{cases} a x^3 - c x - m = 0 & (37\,a) \\ x_{k+1} = \dfrac{2\sqrt{c}}{\sqrt{3\,a}}\cos\left(\dfrac{1}{3}\cos^{-1}\left(\dfrac{3\sqrt{3}\,a\,m}{2\,c\sqrt{c}}\right) + \dfrac{2\pi k}{3}\right), k = 0,1,2. & (37\,b) \end{cases} \quad (37)$$

7. Next, we want to have bx² term appearing on equation (37a). By substituting x with $x + \dfrac{b}{3a}$ into (37) (which means both (37a) and (37b)), then expanding and collecting the terms, which gives

$$\begin{cases} a x^3 + b x^2 + \left(\dfrac{b^2}{3a} - c\right)x + \dfrac{b^3}{27\,a^2} - \dfrac{bc}{3a} - m = 0 & (38\,a) \\ x_{k+1} + \dfrac{b}{3a} = \dfrac{2\sqrt{c}}{\sqrt{3\,a}}\cos\left(\dfrac{1}{3}\cos^{-1}\left(\dfrac{3\sqrt{3}\,a\,m}{2\,c\sqrt{c}}\right) + \dfrac{2\pi k}{3}\right), k = 0,1,2. & (38\,b) \end{cases} \quad (38)$$

8. Replacing c with $\dfrac{b^2}{3a} - c$ for a ≠ 0 into (38) gives

$$\begin{cases} a x^3 + b x^2 + c x + \dfrac{b^3}{27\,a^2} - \dfrac{b}{3a}\left(\dfrac{b^2}{3a} - c\right) - m = 0 & (39\,a) \\ x_{k+1} = -\dfrac{b}{3a} + \dfrac{2}{3a}\sqrt{b^2 - 3\,a\,c}\cdot\cos\left(\dfrac{1}{3}\cos^{-1}\left(\dfrac{27\,a^2\,m}{2\left(b^2 - 3\,a\,c\right)^{3/2}}\right) + \dfrac{2\pi k}{3}\right), k = 0,1,2. & (39\,b) \end{cases} \quad (39)$$

9. Next, we want to have d term appearing on (39a) by setting

$$\dfrac{b^3}{27\,a^2} - \dfrac{b}{3a}\left(\dfrac{b^2}{3a} - c\right) - m = d,$$

and solving for m, which gives

$$m = \dfrac{b^3}{27\,a^2} - \dfrac{b}{9\,a^2}\left(b^2 - 3\,a\,c\right) - d$$

$$= \dfrac{9\,a\,b\,c - 2b^3 - 27\,a^2\,d}{27\,a^2}.$$

10. By substituting m into (39) and rearranging the terms, we obtain the solution of the cubic equation, namely

$$\begin{cases} a x^3 + b x^2 + c x + d = 0 \quad (a \neq 0) & (40\,a) \\ x_{k+1} = -\dfrac{b}{3a} + \dfrac{2}{3a}\sqrt{b^2 - 3ac} \cdot \cos\left(\dfrac{1}{3}\cos^{-1}\left(\dfrac{9abc - 2b^3 - 27a^2 d}{2\left(b^2 - 3ac\right)^{3/2}}\right) + \dfrac{2\pi k}{3}\right), \; k = 0,1,2, & (40\,b) \end{cases} \quad (40)$$

where

$$b^2 - 3ac > 0,$$

$$\left|\frac{9abc - 2b^3 - 27a^2 d}{2\left(b^2 - 3ac\right)^{3/2}}\right| \leq 1.$$

Note: The domain of inverse cosine, $\cos^{-1}(x)$, is restricted to $[-1,1]$. We will provide the analysis of expressions ($9abc-2b^3-27a^2d$) and ($b-3ac$) in (40b) in the summary section.

Example. We show an example of using cubic formula (40b) to solve the cubic equation: $x^3 - 6x^2 + 11x - 6 = 0$.

Solution. The coefficients for the given cubic equation are a = 1, b = -6, c = 11, and d = -6.

Evaluating $\dfrac{9abc - 2b^3 - 27a^2 d}{2\left(b^2 - 3ac\right)^{3/2}} = \dfrac{9\cdot 1 \cdot (-6)\cdot(11) - 2(-6)^3 - 27\cdot 1^2\cdot(-6)}{2\left((-6)^2 - 3\cdot 1\cdot(11)\right)^{3/2}} = 0,$

which belongs to the interval [-1, 1]. Therefore, by applying cubic formula (40b) and substituting the coefficients, which gives the solution to the given equation

$$x_{k+1} = 2 + \frac{2}{\sqrt{3}}\cos\left(\frac{1}{3}\cos^{-1}(0) + \frac{2\pi k}{3}\right), k = 0, 1, 2.$$

$$= 2 + \frac{2}{\sqrt{3}}\cos\left(\frac{\pi}{6} + \frac{2\pi k}{3}\right), k = 0, 1, 2.$$

By putting k = 0, 1 and 2, we obtain $x_1 = 3, x_2 = 1, x_3 = 2$. To verify these roots, we substitute values of x₁ ,x₂ ,x₃ into the given equation, resulting in an evaluation that equals zero.

Formula (40b) provides a solution to cubic equation (40a) but this solution does not cover if $\left|\dfrac{9abc-2b^3-27a^2d}{2(b^2-3ac)^{3/2}}\right|>1.$ To find solution in the interval $(-\infty, -1]$ or $[1, \infty)$, we employ the hyperbolic triple angle cosine identity, $\cosh(3x)=4\cosh^2(x)-3\cosh(x)$.

By replacing x with $\cosh^{-1}(x)$, which gives

$$\cosh\left(3\cosh^{-1}(x)\right)=4x^3-3x. \tag{41}$$

The expressions on the right-hand side of formulae (41) and (31) appear to be identical. Therefore, we get analogous solution (40a) to cubic equation (40b) but it contains the hyperbolic functions instead. As a result, we obtain:

$$\begin{cases} ax^3+bx^2+cx+d=0 \quad (a\neq0) & (42a) \\ x_{k+1}=-\dfrac{b}{3a}+\dfrac{2}{3a}\sqrt{b^2-3ac}\cdot\cosh\left(\dfrac{1}{3}\cosh^{-1}\left(\dfrac{9abc-2b^3-27a^2d}{2(b^2-3ac)^{3/2}}\right)+\dfrac{i2\pi k}{3}\right), \ k=0,1,2, & (42b) \end{cases} \tag{42}$$

where i is the imaginary unit.

Formula (42b) provides a solution to cubic equation (42a) in terms of the hyperbolic cosine and its inverse, where the hyperbolic cosine has a periodicity of $2\pi i$. In fact, finding solution in (40b) also implies finding solution in (42b), and vice versa [for proof, see Appendix 2-A-6]. Although the domain of the inverse hyperbolic cosine in (42b) is restricted to $[1, \infty)$, we can extend or mirror such that it has domain $(-\infty, -1]$ using the identity $\cosh^{-1}(x) = i\pi - \cosh^{-1}(x)$ when $x\leq-1$. This leads formula (42b) providing complex roots. Now, we will focus on the analysis to show that both (40b) or (42b) can be used to provide a complete solution to the cubic equation. Another word, we provide a comprehensive coverage across all domains in order to consolidate all possible scenarios becomes pivotal in presenting a complete set of solutions for the cubic equation, as detailed in the following special cases.

Special Cases of (b² – 3ac) in (42b)

(a) Case b²-3ac > 0:

- If $\dfrac{9abc-2b^3-27a^2d}{2(b^2-3ac)^{3/2}}\geq1,$ the equation has one real root and two complex conjugate roots. The solution is given by formula (42b).

- If $\dfrac{9abc-2b^3-27a^2d}{2\left(b^2-3ac\right)^{3/2}}\leq-1$, the equation has one real root and two complex

conjugate roots, given by formula (42b). However, we cannot directly apply it to real numbers within this interval due to the hyperbolic cosine's definition being restricted to the half closed interval [1, ∞). We can derive an equivalent formula explicitly to the other half closed interval (-∞, -1] by utilizing absolute value notation and applying identity (B), which gives

$$x_{k+1}=-\frac{b}{3a}+\frac{2}{3a}\sqrt{b^2-3ac}\cdot\cosh\left(\frac{1}{3}\cosh^{-1}\left(\left|\frac{9abc-2b^3-27a^2d}{2\left(b^2-3ac\right)^{3/2}}\right|\right)+\frac{i\pi(2k+1)}{3}\right),\, k=0,1,2. \qquad (42c)$$

Proof

Since $\dfrac{9abc-2b^3-27a^2d}{2\left(b^2-3ac\right)^{3/2}}<-1,$ formula (42b) is written as

$$x_{j+1}=-\frac{b}{3a}+\frac{2}{3a}\sqrt{b^2-3ac}\cdot\cosh\left(\frac{1}{3}\cosh^{-1}\left(-\left|\frac{9abc-2b^3-27a^2d}{2\left(b^2-3ac\right)^{3/2}}\right|\right)+\frac{i2\pi j}{3}\right),\, j\in\mathbb{Z}.$$

Applying identity (B) gives

$$x_{j+1}=-\frac{b}{3a}+\frac{2}{3a}\sqrt{b^2-3ac}\cdot\cosh\left(\frac{\pi i}{3}-\frac{1}{3}\cosh^{-1}\left(\left|\frac{9abc-2b^3-27a^2d}{2\left(b^2-3ac\right)^{3/2}}\right|\right)+\frac{i2\pi j}{3}\right),\, j\in\mathbb{Z}.$$

$$=-\frac{b}{3a}+\frac{2}{3a}\sqrt{b^2-3ac}\,\cosh\left(-\frac{1}{3}\cosh^{-1}\left(\left|\frac{9abc-2b^3-27a^2d}{2\left(b^2-3ac\right)^{3/2}}\right|\right)+\frac{i\pi(2j+1)}{3}\right),\, j\in\mathbb{Z}.$$

Since the hyperbolic cosine function is even, it can be rewritten as

$$x_{j+1}=-\frac{b}{3a}+\frac{2}{3a}\sqrt{b^2-3ac}\,\cosh\left(\frac{1}{3}\cosh^{-1}\left(\left|\frac{9abc-2b^3-27a^2d}{2\left(b^2-3ac\right)^{3/2}}\right|\right)-\frac{i\pi(2j+1)}{3}\right),\, j\in\mathbb{Z}.$$

The desired result follows after indexing the index j to k, resulting in (42c).

(b) Case b²-3ac < 0:

In this case formula (42b) has the imaginary unit i appearing inside the inverse hyperbolic cosine function that may appear as a complex expression at first glance while coefficients a, b, c and d are still real numbers. However, it still gives one real root and two complex conjugate roots. Alternatively, formula (42b) can be transformed into another form:

$$x_{k+1}=-\frac{b}{3a}-\frac{2}{3a}\sqrt{|b^2-3ac|}\sinh\left(\frac{1}{3}\sinh^{-1}\left(\frac{9abc-2b^3-27a^2d}{2(|b^2-3ac|)^{3/2}}\right)+\frac{i\pi(2k-1)}{3}\right),\ k=0,1,2.\ (42d)$$

Note that a real root is found if k = 2, namely

$$x_3=-\frac{b}{3a}+\frac{2}{3a}\sqrt{|b^2-3ac|}\sinh\left(\frac{1}{3}\sinh^{-1}\left(\frac{9abc-2b^3-27a^2d}{2(|b^2-3ac|)^{3/2}}\right)\right).\qquad (42e)$$

Proof

Since b^2 -3ac < 0, the radical inside formula (42b) involves with complex form. In this case, we can express the square root using the absolute value notation with the imaginary unit i, and formula (42b) is rewritten as

$$x_{k+1}=-\frac{b}{3a}+\frac{2i}{3a}\sqrt{|b^2-3ac|}\cosh\left(\frac{1}{3}\cosh^{-1}\left(i*\frac{9abc-2b^3-27a^2d}{2(|b^2-3ac|)^{3/2}}\right)+\frac{i2\pi k}{3}\right),\ k\in\mathbb{Z}.$$

By applying the formula $\cosh^{-1}(iu)=\frac{i\pi}{2}+\sinh^{-1}(u)$, we can convert the hyperbolic cosine to the hyperbolic sine, which gives

$$x_{k+1}=-\frac{b}{3a}+\frac{2i}{3a}\sqrt{|b^2-3ac|}\cosh\left(\frac{\pi i}{6}+\frac{1}{3}\sinh^{-1}\left(\frac{9abc-2b^3-27a^2d}{2(|b^2-3ac|)^{3/2}}\right)+\frac{i2\pi k}{3}\right),\ k\in\mathbb{Z}.$$

Since (iπ)/6 can be expressed as (iπ)/2 – (iπ)/3, substituting this expression and rearranging the terms yields:

$$x_{k+1}=-\frac{b}{3a}+\frac{2i}{3a}\sqrt{|b^2-3ac|}\cosh\left(\frac{\pi i}{2}+\frac{1}{3}\sinh^{-1}\left(\frac{9abc-2b^3-27a^2d}{2(|b^2-3ac|)^{3/2}}\right)+\frac{i\pi(2k-1)}{3}\right),\ k\in\mathbb{Z}.$$

By applying formula (C), we obtain the desired result.

(c) Case b^2 -3ac = 0:

We observe that (42) can give solution if $b^2-3ac=0$ or $b=\pm\sqrt{3ac}$. Indeed, the roots can be derived from (42b) by taking the limit of b such that b approaches $\pm\sqrt{3ac}$, and it gives

$$x_{k+1}=-\frac{\pm\sqrt{3\,a\,c}}{3\,a}+\frac{1}{3\,a}\sqrt[3]{\pm(3\,a\,c)^{3/2}-27\,a^2\,d}\;e^{(2i\pi k/3)},\;k=0,1,2. \qquad (43)$$

The choice of $\pm$ depends on the sign of b. If b $\geq$ 0, the positive sign is used; whereas the negative sign is used. Formula (43) gives one real root and two complex conjugate roots. It also gives a triple root if expression $\pm(3ac)^{3/2}-27a^2d=0$. Formula (43) can be expressed in terms of b,

$$x_{k+1}=-\frac{b}{3a}+\frac{1}{3a}\sqrt[3]{b^3-27\,a^2\,d}\;e^{(2i\pi k/3)},\;k=0,1,2. \qquad (44)$$

Proof

To prove (43), we define

$$u=b^2-3a\,c,$$

$$A=\frac{2}{3\,a},$$

$$B=\lim_{b\to\pm\sqrt{ac}}\frac{9\,a\,b\,c-2\,b^3-27\,a^2\,d}{2}=\frac{\pm(3\,a\,c)^{3/2}-27\,a^2\,d}{2},$$

then (42b) becomes

$$x_{k+1}=-\frac{b}{3a}+A\sqrt{u}\cosh\left(\frac{1}{3}\cosh^{-1}\left(\frac{B}{u^{3/2}}\right)+\frac{i\,2\,\pi\,k}{3}\right),\;k=0,1,2.$$

We now consider taking the limit of the expression, $A\sqrt{u}\cosh\left(\frac{1}{3}\cosh^{-1}\frac{B}{u^{3/2}}\right)$, as u tends to 0 as follows:

$$\lim_{u\to0}A\sqrt{u}\cosh\left(\frac{1}{3}\cosh^{-1}\frac{B}{u^{3/2}}\right)=\lim_{u\to0}A\sqrt{u}\cosh\left(\frac{1}{3}\ln\left(\frac{B}{u^{3/2}}+\sqrt{\frac{B^2}{u^3}-1}\right)\right)$$

$$=\lim_{u\to0}A\sqrt{u}\cosh\left(\frac{1}{3}\ln\left(\frac{B}{u^{3/2}}+\frac{1}{u^{3/2}}\sqrt{B^2-u^3}\right)\right)$$

$$=\lim_{u\to0}A\sqrt{u}\cosh\left(\frac{1}{3}\ln\left(B+\sqrt{B^2-u^3}\right)-\frac{1}{2}\ln u\right)$$

$$= \lim_{u \to 0} A\sqrt{u} \cosh\left(\frac{1}{3}\ln\left(B+\sqrt{B^2-u^3}\right)\right)\cosh\left(\frac{1}{2}\ln u\right) -$$

$$A\sqrt{u}\sinh\left(\frac{1}{3}\ln\left(B+\sqrt{B^2-u^3}\right)\right)\sinh\left(\frac{1}{2}\ln u\right)$$

(Using results [2-A-7]: $\lim\limits_{x\to 0}\sqrt{x}\cosh\left(\frac{1}{2}\ln(x)\right)=\frac{1}{2}$, and $\lim\limits_{x\to 0}\sqrt{x}\sinh\left(\frac{1}{2}\ln(x)\right)=-\frac{1}{2}$.)

$$= \frac{A}{2}\cosh\left(\frac{1}{3}\ln(2B)\right)+\frac{A}{2}\sinh\left(\frac{1}{3}\ln(2B)\right)$$

$$= \frac{A}{2}\left[\cosh\left(\frac{1}{3}\ln(2B)\right)+\sinh\left(\frac{1}{3}\ln(2B)\right)\right]$$

$$= \frac{A}{2}\exp\left(\frac{1}{3}\ln(2B)+\frac{2\pi i k}{3}\right)$$

$$= \frac{A\sqrt[3]{2B}}{2}\exp\left(\frac{2\pi i k}{3}\right) \tag{45}$$

Substituting A and B into (45) yields the desired result (43).

Example. Use cubic formula to find the roots of the cubic equation $x^3+3x^2+3x+5=0$.

Solution. Identify the coefficients: a = 1, b = 3, c = 3, and d = 5. Since b = 3 > 0 and $b^2-3ac=3^2-(3)(1)(3) = 0$, we use cubic formula (43), which is

$$x_{k+1}=-\frac{\sqrt{3ac}}{3a}+\frac{1}{3a}\sqrt[3]{(3ac)^{3/2}-27a^2 d}\, e^{(2i\pi k/3)}, \quad k=0,1,2.$$

Substituting the coefficients gives

$$x_{k+1}=-1+\sqrt[3]{-4}\, e^{i2\pi k/3}, \quad k=0,1,2.$$

For k = 0, 1 and 2, the roots to the given equation are:

$$x_1=-1-\sqrt[3]{4},$$

$$x_2=-1+\frac{1}{\sqrt[3]{2}}-\frac{i\sqrt{3}}{\sqrt[3]{2}}$$

and

$$x_3 = -1 + \frac{1}{\sqrt[3]{2}} + \frac{i\sqrt{3}}{\sqrt[3]{2}}.$$

Example. Using the cubic formula to find the roots of the cubic equation,

$x^3 - 3x^2 + 3x + 4 = 0$.

Solution. The coefficients of the given equation are a = 1, b = -3, c = 3, and d = 5. Since b = -3 < 0 and $b^2 - 3ac = 3 - 3(1)(3) = 0$, we use cubic formula (43),

$$x_{k+1} = \frac{\sqrt{3ac}}{3a} + \frac{1}{3a} \sqrt[3]{-(3ac)^{3/2} - 27a^2 d\, e^{(2i\pi k/3)}}, \quad k = 0,1,2.$$

(Notice that the negative sign is used if b < 0.)

By substitution the coefficients, the roots to the given equation are simplified as

$$x_{k+1} = 1 + \sqrt[3]{-5}\, e^{i2\pi k/3}, \quad k = 0,1,2.$$

Then for k = 0, 1 and 2, we get:

$$x_1 = 1 - \sqrt[3]{5},$$

$$x_2 = 1 + \frac{\sqrt[3]{5}}{2} - \frac{i}{2}\sqrt{3}\sqrt[3]{5}$$

and

$$x_3 = 1 + \frac{\sqrt[3]{5}}{2} + \frac{i}{2}\sqrt{3}\sqrt[3]{5}.$$

Before we provide a summary of solution to the cubic equation, we need a transformative tool to transform solution formulas in radicals. As we've explored the quadratic and cubic formulas expressed in trigonometric or hyperbolic trigonometric functions, it's important to note that we haven't addressed how to represent them in radical forms. Let's take a brief pause to discuss another valuable tool specifically designed to transform solution formulas expressed in the forms of trigonometric or hyperbolic trigonometric functions into radical forms.

Transforming Solution Formulas from Trigonometric or Hyperbolic Trigonometric Functions into Radical Forms

As demonstrated in the previous sections, we have derived solution formulas for quadratic and cubic equations expressed in trigonometric or hyperbolic trigonometric functions, namely $\cos((1/n)\cos^{-1}(x))$, $\cosh((1/n)\cosh^{-1}(x))$, $\sin((1/n)\sin^{-1}(x))$, $\sinh((1/n)\sinh^{-1}(x))$, $\tan((1/n)\tan^{-1}(x))$ or $\tanh((1/n)\tanh^{-1}(x))$. However, these solutions are not yet expressed in radical forms. To achieve this, we need to develop a general tool capable of transforming solution formulas expressed in trigonometric or hyperbolic trigonometric functions into radical forms, ensuring that the roots remain the same.

a. Hyperbolic cosine and its inverse

To convert solution formula expressed in hyperbolic cosine to radical form, we use general hyperbolic cosine formula (1.3.3a) and expand it as follows:

$$\cosh\left(\frac{1}{n}\cosh^{-1}(x)+\frac{i2\pi k}{n}\right)=\cosh\left(\frac{1}{n}\ln\left(x+\sqrt{x^2-1}\right)+\frac{i2\pi k}{n}\right)$$

$$=\cosh\left(\ln\left(x+\sqrt{x^2-1}\right)^{1/n}+\frac{i2\pi k}{n}\right)$$

$$=\cosh\left(\ln\left(x+\sqrt{x^2-1}\right)^{1/n}\right)\cosh\left(\frac{i2\pi k}{n}\right)+\sinh\left(\ln\left(x+\sqrt{x^2-1}\right)^{1/n}\right)\sinh\left(\frac{i2\pi k}{n}\right)$$

$$=\left[\frac{1}{2}\left(x+\sqrt{x^2-1}\right)^{1/n}+\frac{1}{2}\left(x+\sqrt{x^2-1}\right)^{-1/n}\right]\frac{1}{2}\left(e^{i2\pi k/n}+e^{-i2\pi k/n}\right)$$

$$+\left[\frac{1}{2}\left(x+\sqrt{x^2-1}\right)^{1/n}-\frac{1}{2}\left(x+\sqrt{x^2-1}\right)^{-1/n}\right]\frac{1}{2}\left(e^{i2\pi k/n}-e^{-i2\pi k/n}\right)$$

$$=\frac{1}{2}\left(x+\sqrt{x^2-1}\right)^{1/n}e^{i2\pi k/n}+\frac{1}{2}\left(x+\sqrt{x^2-1}\right)^{-1/n}e^{-i2\pi k/n}$$

Hence, the radical form of the hyperbolic cosine and its inverse is

$$\cosh\left(\frac{1}{n}\cosh^{-1}(x)+\frac{i2\pi k}{n}\right)=\frac{1}{2}\left[\left(x+\sqrt{x^2-1}\right)^{1/n}e^{i2\pi k/n}+\left(x+\sqrt{x^2-1}\right)^{-1/n}e^{-i2\pi k/n}\right],\begin{array}{l}1\leq x<\infty,\\n\in\mathbb{N},k\in\mathbb{Z}.\end{array}\quad(46a)$$

Formula (46a) represents the radical form for an nth degree equation. The roots expressed in this radical form identical match those derived from the original expression involving hyperbolic cosine and its inverse corresponding to k = 0,1,2, and so on. This exemplifies mathematical beauty and holds significant importance in finding nth roots of equations.

By using a similar approach as described above, we can derive formulas for other functions related to sinh, tanh, coth, sin, tan and cot. The results of these functions are listed below.

b. Hyperbolic sine and its inverse

$$\sinh\left(\frac{1}{n}\sinh^{-1}(x)+\frac{i2\pi k}{n}\right)=\frac{1}{2}\left[\left(x+\sqrt{x^2+1}\right)^{1/n}e^{i2\pi k/n}-\left(x+\sqrt{x^2+1}\right)^{-1/n}e^{-i2\pi k/n}\right],\ \begin{array}{l}1\le x<\infty,\\ n\in\mathbb{N},k\in\mathbb{Z}.\end{array}\quad(46b)$$

Notice that $\left(x+\sqrt{x^2+1}\right)^{-1/n}=\left(\sqrt{x^2-1}-x\right)^{1/n}$, hence we can express (46b) in the alternative form,

$$\sinh\left(\frac{1}{n}\sinh^{-1}(x)+\frac{i2\pi k}{n}\right)=\frac{1}{2}\left[\left(\sqrt{x^2+1}+x\right)^{1/n}e^{i2\pi k/n}-\left(\sqrt{x^2+1}-x\right)^{1/n}e^{-i2\pi k/n}\right],\ \begin{array}{l}1\le x<\infty,\\ n\in\mathbb{N},k\in\mathbb{Z}.\end{array}\quad(46*b)$$

c. Hyperbolic tangent and its inverse

$$\tanh\left(\frac{1}{n}\tanh^{-1}(x)+\frac{i\pi k}{n}\right)=\frac{(1+x)^{1/n}e^{2i\pi k/n}-(1-x)^{1/n}}{(1+x)^{1/n}e^{2i\pi k/n}+(1-x)^{1/n}}\quad -1<x<1\qquad(46c)$$

d. Hyperbolic cotangent and its inverse

$$\coth\left(\frac{1}{n}\coth^{-1}(x)+\frac{i\pi k}{n}\right)=\frac{(x+1)^{1/n}e^{2i\pi k/n}+(x-1)^{1/n}}{(x+1)^{1/n}e^{2i\pi k/n}-(x-1)^{1/n}}\quad -\infty<x<-1\,\text{or}\,1<x<\infty\qquad(46d)$$

e. Cosine and its inverse

$$\cos\left(\frac{1}{n}\cos^{-1}(x)+\frac{2\pi k}{n}\right)=\frac{1}{2}\left[\left(x-i\sqrt{1-x^2}\right)^{1/n}e^{-i2\pi k/n}+\left(x-i\sqrt{1-x^2}\right)^{-1/n}e^{i2\pi k/n}\right]\ -1\le x\le1\quad(46e)$$

$$\cos\left(\frac{1}{n}\cos^{-1}(x)-\frac{\pi(2k+1)}{n}\right)=\frac{1}{2}\left[\left(-x+i\sqrt{1-x^2}\right)^{1/n}e^{i2\pi k/n}+\left(-x+i\sqrt{1-x^2}\right)^{-1/n}e^{-i2\pi k/n}\right]\ -1\le x\le1$$

f. Sine and its inverse

$$\sin\left(\frac{1}{n}\sin^{-1}(x)+\frac{2\pi k}{n}\right)=\frac{1}{2i}\left[\left(ix+\sqrt{1-x^2}\right)^{1/n}e^{i2\pi k/n}-\left(ix+\sqrt{1-x^2}\right)^{-1/n}e^{-i2\pi k/n}\right]\ -1\le x\le1\quad(46f)$$

g. Tangent and its inverse

$$\tan\left(\frac{1}{n}\tan^{-1}(x)+\frac{\pi k}{n}\right)=\frac{1}{i}\frac{(1+ix)^{1/n}e^{2i\pi k/n}-(1-ix)^{1/n}}{(1+ix)^{1/n}e^{2i\pi k/n}+(1-ix)^{1/n}}\quad -\infty<x<\infty\qquad(46g)$$

h. Cotangent and its inverse

$$\cot\left(\frac{1}{n}\cot^{-1}(x)-\frac{\pi k}{n}\right)=\begin{cases} i\dfrac{(ix+1)^{1/n}e^{2i\pi k/n}+(ix-1)^{1/n}}{(ix+1)^{1/n}e^{2i\pi k/n}-(ix-1)^{1/n}} & x>0 \\[4mm] -i\dfrac{(ix+1)^{1/n}e^{2i\pi k/n}+(ix-1)^{1/n}}{(ix+1)^{1/n}e^{2i\pi k/n}-(ix-1)^{1/n}} & x<0 \end{cases}$$

(46 h)

Furthermore, the solution formulas for the nth degree equation can be expressed in the radical forms by employing various derived expressions, such as (1.10a) for cosine, (1.2.9) for hyperbolic cosine, (1.13.7a) or (1.13.7b) for tangent, (1.13.15) for hyperbolic tangent, and so forth, as outlined in Chapter 1.

Let us now return to our discussion regarding the analysis of the solution formulas for the cubic equation.

Summary of Solution to Cubic Equation (Type 1)

Cubic equation: $a x^3+b x^2+c x+d=0,\quad (a\neq 0)$

Defining

$$D=b^2-3ac \tag{47 a}$$

$$T=\frac{9abc-2b^3-27a^2d}{2D^{3/2}} \tag{47 b}$$

Case D > 0:

> If T ≥ 1, the equation has one real root and two complex conjugate roots,

$$x_{k+1}=-\frac{b}{3a}+\frac{2}{3a}\sqrt{D}\cosh\left(\frac{1}{3}\cosh^{-1}(T)+\frac{i2\pi k}{3}\right),\, k=0,1,2. \tag{47 c}$$

(Notice that the formula is still valid if T = 1.)

Applying (46a) gives a radical form:

$$x_{k+1}=-\frac{b}{3a}+\frac{\sqrt{D}}{3a}\cdot\left[\left(T+\sqrt{T^2-1}\right)^{1/3}e^{i2\pi k/3}+\left(T+\sqrt{T^2-1}\right)^{-1/3}e^{-i2\pi k/3}\right] \tag{47 d}$$

> If T ≤ -1, the equation has one real root and two complex conjugate roots,

which are given by

$$x_{k+1}=-\frac{b}{3a}+\frac{2\sqrt{D}}{3a}\cosh\left(\frac{1}{3}\cosh^{-1}(|T|)+\frac{i\pi(2k+1)}{3}\right), k=0,1,2.\qquad(47e)$$

Using (46a), formula (47e) is converted to radical form:

$$x_{k+1}=-\frac{b}{3a}+\frac{\sqrt{D}}{3a}\cdot\left[\left(|T|+\sqrt{T^2-1}\right)^{1/3}e^{i\pi(2k+1)/3}+\left(|T|+\sqrt{T^2-1}\right)^{-1/3}e^{-i\pi(2k+1)/3}\right]\qquad(47f)$$

Note that the real root is found if k = 1, namely

$$x_3=-\frac{b}{3a}-\frac{2}{3a}\sqrt{D}\cosh\left(\frac{1}{3}\cosh^{-1}(|T|)\right).\qquad(47f)$$

> If -1≤T≤1, then the equation has three distinct real roots

$$x_{k+1}=-\frac{b}{3a}+\frac{2}{3a}\sqrt{D}\cos\left(\frac{1}{3}\cos^{-1}(T)+\frac{2\pi k}{3}\right), k=0,1,2.\qquad(47g)$$

Formula (47g) is transformed into the radical form using (46e):

$$x_{k+1}=-\frac{b}{3a}+\frac{\sqrt{D}}{3a}\cdot\left[\left(T-i\sqrt{1-T^2}\right)^{1/3}e^{-i2\pi k/3}+\left(T-i\sqrt{1-T^2}\right)^{-1/3}e^{i2\pi k/3}\right], k=0,1,2.\quad(47h)$$

Case D < 0:

> In this case, T becomes

$$T=\frac{9abc-2b^3-27a^2d}{2|D|^{3/2}}.$$

The equation has one real root and two complex conjugate roots, which are given by (42d):

$$x_{k+1}=-\frac{b}{3a}-\frac{2}{3a}\sqrt{|D|}\sinh\left(\frac{1}{3}\sinh^{-1}(T)+\frac{i\pi(2k-1)}{3}\right), k=0,1,2.\qquad(47i)$$

Note that the real root is found if k = 2, namely

$$x_3=-\frac{b}{3a}+\frac{2}{3a}\sqrt{|D|}\sinh\left(\frac{1}{3}\sinh^{-1}(T)\right)\qquad(47k)$$

and solution (47i) can be expressed in the radical form as

$$x_{k+1}=-\frac{b}{3a}-\frac{\sqrt{|D|}}{3a}\cdot\left[\left(T+\sqrt{T^2+1}\right)^{1/3}e^{i\pi(2k-1)/3}-\left(T+\sqrt{T^2+1}\right)^{-1/3}e^{-i\pi(2k-1)/3}\right],k=0,1,2. \quad (47l)$$

or in the alternative form,

$$x_{k+1}=-\frac{b}{3a}+\frac{\sqrt{|D|}}{3a}\cdot\left[\left(T+\sqrt{T^2+1}\right)^{1/3}e^{i2\pi k/3}-\left(T+\sqrt{T^2+1}\right)^{-1/3}e^{-i2\pi k/3}\right],k=0,1,2. \quad (47l)$$

Case D = 0:

➤ In this case, the equation has one real root and two complex conjugate roots, which are given by (43) or (44):

$$x_{k+1}=-\frac{\pm\sqrt{3ac}}{3a}+\frac{1}{3a}\sqrt[3]{\pm(3ac)^{3/2}-27a^2d}\,e^{(2i\pi k/3)},\quad k=0,1,2, \quad (47m)$$

where the choice of ± depends on the sign of b. The positive sign is used if b ≥ 0; otherwise, the negative sign is used. The solution can be expressed in terms of b,

$$x_{k+1}=-\frac{b}{3a}+\frac{1}{3a}\sqrt[3]{b^3-27a^2d}\,e^{(2i\pi k/3)},\quad k=0,1,2. \quad (47n)$$

Note:

$$e^{i2\pi/3}=\frac{1}{2}\left(-1+i\sqrt{3}\right) \text{ and } e^{i4\pi/3}=\frac{1}{2}\left(-1-i\sqrt{3}\right). \quad (47o)$$

➤ If $b=c=3\sqrt[3]{ad^2}$, the equation has a triple root,

$$\begin{cases} x_1=x_2=x_3=-\sqrt[3]{\dfrac{d}{a}}, & b\geq 0 \\[2mm] x_1=x_2=x_3=\sqrt[3]{\dfrac{d}{a}}, & b<0. \end{cases} \quad (47p)$$

Notes

1. i is the imaginary unit.

2. Formulas (47d), (47f), (47h) or (47l) can be further simplified by removing the expression D from the denominator when we substitute T into these formulas.

Cubic Discriminant

We observe that the result

$$\left(2\left(b^2-3ac\right)^{3/2}\right)^2-\left(9abc-2b^3-27a^2d\right)^2=27a^2\left(27a^2d^2+b^2c^2+18abcd-4b^3d-4ac^2\right), \quad (47n)$$

in which the expression $27a^2d^2+b^2c^2+18abcd-4b^3d-4ac^3$ is a widely recognized *discriminant* of the cubic equation denoted D₂, a term initially coined as the *"discriminant"* [10] by James Joseph Sylvester in 1851. Therefore, T can be expressed in terms of D₂ as follows:

$$T^2=1-\frac{27a^2D_2}{4\left(b^2-3ac\right)^3}. \quad\quad (47o)$$

(Since D notation is already used earlier, D₂ temporarily used in this note represent the discriminate of the cubic equation.)

How to Use the Cubic Formula in Practice

Use *Summary of Solution to Cubic Equation* and its *Notes*, and follow these steps to apply the cubic formula and find solution to cubic equation (30):

- Determine the values of D and T.

- If D > 0 and -1≤T≤1, formula (40b) or (47f) involving with the cosine is applied. Otherwise, the formulas with the hyperbolic cosine function is used: i. If T≥1, formula (42b) or (47c) involving the hyperbolic cosine is applied. ii. If T < -1, formula (42c) or (47d) is applied. These formulas are almost identical and differ only in the trigonometric or hyperbolic trigonometric functions they contain.

- If D = 0, then T is undefined. Formula (43) or (47i) is applied. In this case, the choice of ± depends on the sign of b: The positive sign is used if b ≥ 0; otherwise, the negative sign is used. If $b=c=3\sqrt[3]{ad^2}$, the cubic equation has multiple roots (47m) or triple roots.

Example. Find the roots of the cubic equation $x^3-3x^2+2x-1=0$.

Solution. The coefficients of the given equation are a = 1, b = -3, c = 2, and d = -1.

- The values of D and T are:

$$D = b^2 - 3ac = (-3)^2 - 3(1)(2) = 3 > 0$$

and

$$T = \frac{9abc - 2b^3 - 27a^2d}{2(b^2 - 3ac)^{3/2}} = \frac{9 \cdot 1 \cdot (-2) \cdot 3 - 2 \cdot (-2)^3 - 27 \cdot 1^2 \cdot (-1)}{2((-1)^2 - 3 \cdot a \cdot 3)^{3/2}} = \frac{3\sqrt{3}}{2} > 1.$$

- In this case, we have D > 0 and T > 1, therefore, the cubic formula providing the solution for given equation is:

$$x_{k+1} = -\frac{b}{3a} + \frac{2}{3a}\sqrt{D}\cosh\left(\frac{1}{3}\cosh^{-1}(T) + \frac{i2\pi k}{3}\right), k = 0, 1, 2.$$

- By substituting the coefficients, we obtain:

$$x_1 = 1 + \frac{2}{\sqrt{3}}\cosh\left(\frac{1}{3}\cosh^{-1}\left(\frac{3\sqrt{3}}{2}\right)\right), \qquad \text{(real root)}$$

$$x_2 = 1 + \frac{2}{\sqrt{3}}\cosh\left(\frac{1}{3}\cosh^{-1}\left(\frac{3\sqrt{3}}{2}\right) + \frac{i2\pi}{3}\right) \qquad \text{(complex root)}$$

and

$$x_3 = 1 + \frac{2}{\sqrt{3}}\cosh\left(\frac{1}{3}\cosh^{-1}\left(\frac{3\sqrt{3}}{2}\right) + \frac{i4\pi}{3}\right). \qquad \text{(complex root)}$$

Example. Find the roots of the cubic equation $x^3 - 3x^2 + x + 5 = 0$.

Solution. The coefficients of the given equation are a = 1, b = -3, c = 1, and d = 5. We consider the following steps:

- The values of D and T are:

$$D = b^2 - 3ac = (-3)^2 - 3(1)(1) = 6 > 0$$

and

$$T = \frac{9abc - 2b^3 - 27a^2d}{2D^{3/2}}$$

$$= -\frac{3\sqrt{6}}{2} < -1.$$

- In this case, we have D > 0 and T < -1, the cubic formula providing the solution to the given formula is:

$$x_{k+1} = -\frac{b}{3a} + \frac{2\sqrt{D}}{3a} \cosh\left(\frac{1}{3}\cosh^{-1}(|T|) + \frac{i\pi(2k+1)}{3}\right), k = 0, 1, 2.$$

- By substituting the coefficients, we get:

$$x_1 = 1 + \frac{2\sqrt{6}}{3}\cosh\left(\frac{1}{3}\cosh^{-1}\left(\frac{3\sqrt{6}}{2}\right) + \frac{i\pi}{3}\right),$$

$$x_2 = 1 + \frac{2\sqrt{6}}{3}\cosh\left(\frac{1}{3}\cosh^{-1}\left(\frac{3\sqrt{6}}{2}\right) + i\pi\right) = 1 - \frac{2\sqrt{6}}{3}\cosh\left(\frac{1}{3}\cosh^{-1}\left(\frac{3\sqrt{6}}{2}\right)\right)$$

and

$$x_3 = 1 + \frac{2\sqrt{6}}{3}\cosh\left(\frac{1}{3}\cosh^{-1}\left(\frac{3\sqrt{6}}{2}\right) + \frac{i5\pi}{3}\right).$$

B. Solution to Cubic Equation (Type 2)

In this section, we derive a solution to cubic equation (30) by using the triple angle identity for hyperbolic sine [2-A-8] which is expressed as

$$\sinh(3x) = 4\sinh^3 x + 3\sinh x \tag{47}$$

We do the same procedure as previously described in Part A as shown in the following steps:

1. By multiplying through both sides of (47) by a/4 with a≠0, which gives

$$\frac{a}{4}\sinh(3x) = a\sinh^3 x + \frac{3a}{4}\sinh x$$

2. Replacing x with sinh^{-1}(x) gives

$$\frac{a}{4}\sinh(3\sinh^{-1}(x)) = ax^3 + \frac{3a}{4}x$$

3. Adding m to both sides gives

$$\frac{a}{4}\sinh\left(3\sinh^{-1}(x)\right)+m=a\,x^3+\frac{3\,a}{4}\,x+m \tag{48}$$

4. Setting both sides of (48) to zero which gives a system of two simultaneous equations

$$\begin{cases} a\,x^3+\dfrac{3\,a}{4}\,x+m=0 & (48\,a)\\[3mm] x=-\sinh\left(\dfrac{1}{3}\sinh^{-1}\left(\dfrac{4\,m}{a}\right)\right) & (48\,b) \end{cases}$$

5. Replacing x with x + b/(3a) gives

$$\begin{cases} a\,x^3+b\,x^2+\left(\dfrac{b^2}{3\,a}+\dfrac{3\,a}{4}\right)x+m+\dfrac{b^3}{27\,a^2}+\dfrac{b}{4}=0 & (49\,a)\\[3mm] x=-\dfrac{b}{3\,a}-\sinh\left(\dfrac{1}{3}\sinh^{-1}\left(\dfrac{4\,m}{a}\right)\right) & (49\,b) \end{cases} \tag{49}$$

6. Replacing m with $d-\dfrac{b^3}{27a^2}-\dfrac{b}{4}$ gives

$$\begin{cases} a\,x^3+b\,x^2+\left(\dfrac{b^2}{3\,a}+\dfrac{3\,a}{4}\right)x+d=0 & (50\,a)\\[3mm] x=-\dfrac{b}{3\,a}-\sinh\left(\dfrac{1}{3}\sinh^{-1}\left(\dfrac{108\,a^2\,d-27\,a^2\,b-4\,b^3}{27\,a^3}\right)\right) & (50\,b) \end{cases} \tag{50}$$

7. Replacing x with x/c gives

$$\begin{cases} a\,x^3+b\,c\,x^2+\left(\dfrac{b^2\,c^2}{3\,a}+\dfrac{3\,a\,c^2}{4}\right)x+c^3\,d=0 & (51\,a)\\[3mm] x=-\dfrac{b\,c}{3\,a}-c\sinh\left(\dfrac{1}{3}\sinh^{-1}\left(\dfrac{-27\,a^2\,b+108\,a^2\,d-4\,b^3}{27\,a^3}\right)\right) & (51\,b) \end{cases} \tag{51}$$

8. Replacing b with b/c gives

$$\left|\begin{array}{l} a\,x^3+b\,x^2+\left(\dfrac{b^2}{3\,a}+\dfrac{3\,a\,c^2}{4}\right)x+c^3\,d=0 \qquad\qquad\qquad (52\,a) \\[3mm] x=-\dfrac{b}{3\,a}+c\,\sinh\left(\dfrac{1}{3}\sinh^{-1}\left(\dfrac{27\,a^2\,b\,c^2-108\,a^2\,c^3\,d+4\,b^3}{27\,a^3\,c^3}\right)\right) \qquad (52\,b) \end{array}\right. \qquad (52)$$

9. Replacing d with d/c^3 gives

$$\left|\begin{array}{l} a\,x^3+b\,x^2+\left(\dfrac{b^2}{3\,a}+\dfrac{3\,a\,c^2}{4}\right)x+d=0 \qquad\qquad\qquad (53\,a) \\[3mm] x=-\dfrac{b}{3\,a}+c\,\sinh\left(\dfrac{1}{3}\sinh^{-1}\left(\dfrac{27\,a^2\,b\,c^2-108\,a^2\,d+4\,b^3}{27\,a^3\,c^3}\right)\right) \qquad (53\,a) \end{array}\right. \qquad (53)$$

10. By setting

$$\dfrac{b^2}{3\,a}+\dfrac{3\,a\,c^2}{4}=C \qquad \text{(Note, C temporarily denotes an uppercase coefficient.)}$$

and then solving for c in terms of a, b and C, which yields

$$c=\dfrac{2\sqrt{3\,a\,C-b^2}}{3\,a}.$$

11. By substituting c into (53) and then renaming C as c, which gives

$$\left|\begin{array}{l} a\,x^3+b\,x^2+c\,x+d=0 \quad (a\neq 0) \qquad\qquad\qquad\qquad (54\,a) \\[3mm] x=-\dfrac{b}{3\,a}+\dfrac{2\sqrt{3\,a\,c-b^2}}{3\,a}\sinh\left(\dfrac{1}{3}\sinh^{-1}\left(\dfrac{12\,b\left(3\,a\,c-b^2\right)-108\,a^2\,d+4\,b^3}{8\left(3\,a\,c-b^2\right)^{3/2}}\right)\right) \quad (54\,b) \\[3mm] \text{where } 3\,a\,c-b^2\geq 0. \end{array}\right. \qquad (54)$$

12. After expanding and simplifying the terms in (54b), then expressing its resulting in terms of periodicity of hyperbolic sine, which gives

$$\left|\begin{array}{l} a\,x^3+b\,x^2+c\,x+d=0 \quad (a\neq 0) \qquad\qquad\qquad\qquad (55\,a) \\[3mm] x_{k+1}=-\dfrac{b}{3\,a}+\dfrac{2\sqrt{3\,a\,c-b^2}}{3\,a}\sinh\left(\dfrac{1}{3}\sinh^{-1}\left(\dfrac{9\,a\,b\,c-2\,b^3-27\,a^2\,d}{2\left(3\,a\,c-b^2\right)^{3/2}}\right)+\dfrac{2\,\pi\,i\,k}{3}\right), \quad k=0,1,2. \quad (55\,b) \\[3mm] \text{where } 3\,a\,c-b^2\geq 0. \end{array}\right. \qquad (55)$$

Formula (55b) provides solution to the general cubic equation that cover all real numbers since the domain of inverse hyperbolic sine is defined in the interval (-∞,

∞).

Special Cases of (3ac - b²) in (55b)

➢ If 3ac-b²<0, then the equation has three distinct real roots. Indeed, the solution can be derived from (55b) by applying formula (G) and (F), which gives

$$x_{k+1}=-\frac{b}{3a}-\frac{2}{3a}\sqrt{|3ac-b^2|}\sin\left(\frac{1}{3}\sin^{-1}\left(\frac{9abc-2b^3-27a^2d}{2\left(|3ac-b^2|\right)^{3/2}}\right)+\frac{2\pi k}{3}\right),\quad k=0,1,2,\quad (55c)$$

where $-1\le\dfrac{9abc-2b^3-27a^2d}{2\left(|b^2-3ac|\right)^{3/2}}\le1.$

+> If $\dfrac{9abc-2b^3-27a^2d}{2\left(b^2-3ac\right)^{3/2}}>1$, formula (55c) gives one real root and two complex conjugate roots,

$$x_{k+1}=-\frac{b}{3a}-\frac{2}{3a}\sqrt{|3ac-b^2|}\cosh\left(\frac{1}{3}\cosh^{-1}\left(\frac{9abc-2b^3-27a^2d}{2\left(|3ac-b^2|\right)^{3/2}}\right)+\frac{i2\pi k}{3}\right),\quad k=0,1,2.\ (55d)$$

Proof

We use result (55c) to prove (55d) as follows:

Let $u=\dfrac{9abc-2b^3-27a^2d}{2\left(b^2-3ac\right)^{3/2}}$

Substituting for u in (55c) gives

$$x_{k+1}=-\frac{b}{3a}-\frac{2\sqrt{|3ac-b^2|}}{3a}\sin\left(\frac{1}{3}\sin^{-1}(u)+\frac{2\pi k}{3}\right)$$

$$=-\frac{b}{3a}-\frac{2\sqrt{|3ac-b^2|}}{3a}\sin\left(\frac{1}{3i}\ln\left(iu+\sqrt{1-u^2}\right)+\frac{2\pi k}{3}\right)$$

Suppose that |u| > 1, we have

$$=-\frac{b}{3a}-\frac{2\sqrt{b^2-3ac}}{3a}\sin\left(\frac{1}{3i}\ln\left(iu+i\sqrt{u^2-1}\right)+\frac{2\pi k}{3}\right)$$

Factoring out i from the expression in parentheses gives

$$= -\frac{b}{3a} - \frac{2\sqrt{b^2-3ac}}{3a}\sin\left(\frac{1}{3}\frac{1}{i}\ln(i) + \frac{1}{3i}\ln\left(u+\sqrt{u^2-1}\right) + \frac{2\pi k}{3}\right)$$

Continue substituting $\ln(i)/i = \pi/2$ and $\ln\left(u+\sqrt{u^2-1}\right) = \cosh^{-1}(u)$ yields

$$= -\frac{b}{3a} - \frac{2\sqrt{b^2-3ac}}{3a}\sin\left(\frac{\pi}{6} + \frac{1}{3i}\cosh^{-1}(u) + \frac{2\pi k}{3}\right)$$

Replacing $\pi/6 = \pi/2 - \pi/3$ and rearranging the terms gives

$$= -\frac{b}{3a} - \frac{2\sqrt{b^2-3ac}}{3a}\sin\left(\frac{\pi}{2} + \frac{1}{3i}\cosh^{-1}(u) + \frac{2\pi(k-1)}{3}\right)$$

By applying $\sin(\pi/2 + u) = \cos(u)$, it follows that

$$= -\frac{b}{3a} - \frac{2\sqrt{b^2-3ac}}{3a}\cos\left(\frac{1}{3i}\cosh^{-1}(u) + \frac{2\pi(k-1)}{3}\right)$$

Writing the fractions with a common denominator in parentheses yields

$$= -\frac{b}{3a} - \frac{2\sqrt{b^2-3ac}}{3a}\cos\left(\frac{1}{3i}\cosh^{-1}(u) + \frac{i2\pi(k-1)}{3i}\right)$$

Applying identity (I) gives

$$= -\frac{b}{3a} - \frac{2\sqrt{b^2-3ac}}{3a}\cosh\left(\frac{1}{3}\cosh^{-1}(u) + \frac{i2\pi(k-1)}{3}\right)$$

The result is derived by substituting u for the previously defined expression and then re-indexing the value of k.

+> If $\dfrac{9abc - 2b^3 - 27a^2d}{2\left(b^2-3ac\right)^{3/2}} < -1$, formula (55d) becomes

$$x_{k+1} = -\frac{b}{3a} - \frac{2}{3a}\sqrt{|3ac-b^2|}\cosh\left(\frac{1}{3}\cosh^{-1}\left(-\left|\frac{9abc-2b^3-27a^2d}{2\left(|3ac-b^2|\right)^{3/2}}\right|\right) + \frac{i2\pi k}{3}\right), \quad k=0,1,2.$$

By applying $\cosh^{-1}(-u) = i\pi - \cosh(u)$, which gives one real root and two complex conjugate roots:

$$x_{k+1}=-\frac{b}{3a}-\frac{2}{3a}\sqrt{|3ac-b^2|}\cosh\left(\frac{1}{3}\cosh^{-1}\left(\left|\frac{9abc-2b^3-27a^2d}{2(|3ac-b^2|)^{3/2}}\right|\right)-\frac{i\pi(2k+1)}{3}\right),\ k=0,1,2.\ (55e)$$

➢ If 3ac-b²=0, then formula (55b) becomes

$$x_{k+1}=-\frac{\pm\sqrt{3ac}}{3a}+\frac{1}{3a}\sqrt[3]{\pm(3ac)^{3/2}-27a^2d}\ e^{(2i\pi k/3)},\ k=0,1,2,\qquad(55f)$$

where the choice of ± depends on the sign of b. The positive sign is used if b ≥ 0; otherwise, the negative sign is used. Formula (55f) can be expressed in terms of b,

$$x_{k+1}=-\frac{b}{3a}+\frac{1}{3a}\sqrt[3]{b^3-27a^2d}\ e^{(2i\pi k/3)},\ k=0,1,2.\qquad(55g)$$

Proof
We use (55b) to prove (55f) as follows:

Define $u=3ac-b^2,\quad A=\frac{2}{3a},$

and

$$B=\lim_{b\to\pm\sqrt{ac}}\frac{9abc-2b^3-27a^2d}{2}=\frac{\pm(3ac)^{3/2}-27a^2d}{2}.$$

Then from (55b), we consider taking the limit of the expression below as u tends to 0:

$$\lim_{u\to0}A\sqrt{u}\sinh\left(\frac{1}{3}\cosh^{-1}\frac{B}{u^{3/2}}\right)=\lim_{u\to0}A\sqrt{u}\sinh\left(\frac{1}{3}\ln\left(\frac{B}{u^{3/2}}+\sqrt{\frac{B^2}{u^3}-1}\right)\right)$$

$$=\lim_{x\to0}A\sqrt{u}\sinh\left(\frac{1}{3}\ln\left(\frac{B}{u^{3/2}}+\frac{1}{u^{3/2}}\sqrt{B^2-u^3}\right)\right)$$

$$=\lim_{x\to0}A\sqrt{u}\sinh\left(\frac{1}{3}\ln\left(B+\sqrt{B^2-u^3}\right)-\frac{1}{2}\ln u\right)$$

$$=\lim_{x\to0}A\sqrt{u}\sinh\left(\frac{1}{3}\ln\left(B+\sqrt{B^2-u^3}\right)\right)\cosh\left(\frac{1}{2}\ln u\right)$$

$$-A\sqrt{u}\cosh\left(\frac{1}{3}\ln\left(B+\sqrt{B^2-u^3}\right)\right)\sinh\left(\frac{1}{2}\ln u\right)$$

(Using results [2-A-7] $\lim\limits_{x \to 0} \sqrt{x}\cosh\left(\frac{1}{2}\ln(x)\right)=\frac{1}{2}$, and $\lim\limits_{x \to 0} \sqrt{x}\sinh\left(\frac{1}{2}\ln(x)\right)=-\frac{1}{2}$)

$$= \frac{A}{2}\sinh\left(\frac{1}{3}\ln(2B)\right)+\frac{A}{2}\cosh\left(\frac{1}{3}\ln(2B)\right)$$

$$= \frac{A}{2}\left[\sinh\left(\frac{1}{3}\ln(2B)\right)+\cosh\left(\frac{1}{3}\ln(2B)\right)\right]$$

$$= \frac{A}{2}\exp\left(\frac{1}{3}\ln(2B)+\frac{2\pi i k}{3}\right)$$

$$= \frac{A\sqrt[3]{2B}}{2}\exp\left(\frac{2\pi i k}{3}\right). \tag{56}$$

Substituting A and B into (56) yields the desired result (55e).

Summary of Solution to the Cubic Equation (Type 2)

Cubic equation: $ax^3+bx^2+cx+d=0$ $(a \neq 0)$

Define

$$D=3ac-b^2 \tag{57a}$$

$$T=\frac{9abc-2b^3-27a^2d}{2D^{3/2}} \tag{57b}$$

1. Case D > 0:

The equation has one real root and two complex conjugate roots, given by formula (55b), that can be simplified to the form,

$$x_{k+1}=-\frac{b}{3a}+\frac{2\sqrt{3ac-b^2}}{3a}\sinh\left(\frac{1}{3}\sinh^{-1}(T)+\frac{2\pi i k}{3}\right), \quad k=0,1,2 \tag{57c}$$

and by applying (46b) to (57c), which gives solution in the radical form,

$$x_{k+1}=-\frac{b}{3a}+\frac{\sqrt{|D|}}{3a}\cdot\left[\left(T+\sqrt{T^2+1}\right)^{1/3}e^{i2\pi k/3}-\left(T+\sqrt{T^2+1}\right)^{-1/3}e^{-i2\pi k/3}\right], k=0,1,2. \tag{57d}$$

2. Case D < 0:

In this case, T becomes

$$T = \frac{9\,a\,b\,c - 2\,b^3 - 27\,a^2\,d}{2\,(|D|)^{3/2}} .$$ (57 e)

a. If -1≤T≤1, the equation has three distinct real roots, provided by formula (55c), which can be simplified to the form,

$$x_{k+1} = -\frac{b}{3a} - \frac{2}{3a}\sqrt{|D|}\sin\left(\frac{1}{3}\sin^{-1}(T) + \frac{2\pi k}{3}\right), \quad k = 0, 1, 2.$$ (57 f)

Applying (46f) to (57f) yields the solution in the radical form,

$$x_{k+1} = -\frac{b}{3a} - \frac{\sqrt{|D|}}{i\,3a}\left[\left(iT + \sqrt{1 - T^2}\right)^{1/3} e^{i2\pi k/3} - \left(iT + \sqrt{1 - T^2}\right)^{-1/3} e^{-i2\pi k/3}\right], \quad k = 0, 1, 2.$$ (57 g)

b. If T ≥ 1, the equation has one root and two complex conjugate roots, obtained through formula (55d), which yields to the following hyperbolic and radical forms:

- $$x_{k+1} = -\frac{b}{3a} + \frac{2}{3a}\sqrt{|D|}\cosh\left(\frac{1}{3}\cosh^{-1}(T) + \frac{i2\pi k}{3}\right), \quad k = 0, 1, 2$$ (57 h)

- $$x_{k+1} = -\frac{b}{3a} + \frac{\sqrt{|D|}}{3a}\cdot\left[\left(T + \sqrt{T^2 - 1}\right)^{1/3} e^{i2\pi k/3} + \left(T + \sqrt{T^2 - 1}\right)^{-1/3} e^{-i2\pi k/3}\right].$$ (57 i)

(The radical form can be obtained by applying (46b) to (57).)

c. If T ≤ -1, the equation has one root and two complex conjugate roots, given by formula (55d), which can be simplified as follows:

- Hyperbolic form:

$$x_{k+1} = -\frac{b}{3a} + \frac{2}{3a}\sqrt{|D|}\cosh\left(\frac{1}{3}\cosh^{-1}(|T|) - \frac{i\pi(2k+1)}{3}\right), \quad k = 0, 1, 2.$$ (57 k)

Or

$$x_{k+1} = -\frac{b}{3a} - \frac{2}{3a}\sqrt{|D|}\cosh\left(\frac{1}{3}\cosh^{-1}(|T|) + \frac{i2\pi k}{3}\right), \quad k = 0, 1, 2.$$ (57 l)

- Radical form:

$$x_{k+1} = -\frac{b}{3a} + \frac{\sqrt{|D|}}{3a}\cdot\left[\left(|T| + \sqrt{T^2 - 1}\right)^{1/3} e^{-i\pi(2k+1)/3} + \left(|T| + \sqrt{T^2 - 1}\right)^{-1/3} e^{i\pi(2k+1)/3}\right]$$ (47 m)

3. Case D = 0:

The equation has one real root and two complex conjugate roots, given by (55e) or (55f):

$$x_{k+1} = -\frac{b}{3a} + \frac{1}{3a} \sqrt[3]{b^3 - 27 a^2 d}\, e^{(2i\pi k/3)}, \quad k = 0, 1, 2. \tag{57n}$$

If the expression $\pm(3ac)^{3/2} - 27a^2d = 0$ or $c = 3\sqrt[3]{ad^2}$, then the equation has three identical real roots, namely

$$x_1 = x_2 = x_3 = \pm \sqrt[3]{\frac{d}{a}}, \tag{57o}$$

where the negative sign is used if b is greater than or equal to 0; otherwise, the negative sign is used.

Notes:

- i is the imaginary unit.

- $e^{i2\pi/3} = \frac{1}{2}\left(-1 + i\sqrt{3}\right)$ and $e^{i4\pi/3} = \frac{1}{2}\left(-1 - i\sqrt{3}\right)$.

C. Solution to Cubic Equation (Type 3)

We develop another solution to cubic equation (30) by using the triple angle identity for tangent [2-A-9] which is expressed as

$$\tan(3x) = \frac{3\tan x - \tan^3 x}{1 - 3\tan^2 x} \tag{58}$$

By multiplying through both sides of (58) by a (a ≠ 0), which gives

$$a\tan(3x) = \frac{a\tan^3 x - 3a\tan x}{3\tan^2 x - 1}$$

Then adding b to both sides gives

$$a\tan(3x) + b = \frac{a\tan^3 x - 3a\tan x}{3\tan^2 x - 1} + b$$

By setting both sides to 0, which gives a system of two simultaneous equations:

$$\begin{cases} \dfrac{a\,x^3 - 3\,a\,x}{3\,x^2 - 1} + b = 0 & (a \neq 0) \quad & (58\,a) \\[2mm] a\,\tan\left(3\,\tan^{-1}(x)\right) + b = 0 & & (58\,b) \end{cases}$$

By solving x from (58b) after replacing b by b/3 and multiplying through both sides of (58a) by (3x²-1), we get

$$\begin{cases} a\,x^3 + b\,x^2 - 3\,a\,x - \dfrac{b}{3} = 0 & (a \neq 0) \quad & (59\,a) \\[3mm] x = -\tan\left(\dfrac{1}{3}\tan^{-1}\left(\dfrac{b}{3\,a}\right)\right) & & (59\,b) \end{cases} \quad (59)$$

Replacing x with x/c gives

$$\begin{cases} a\,x^3 + b\,c\,x^2 - 3\,a\,c^2\,x - \dfrac{b\,c^3}{3} = 0 & (a \neq 0) \quad & (60\,a) \\[3mm] x = -c\,\tan\left(\dfrac{1}{3}\tan^{-1}\left(\dfrac{b}{3\,a}\right)\right) & & (60\,a) \end{cases} \quad (60)$$

Replacing b with b/c gives

$$\begin{cases} a\,x^3 + b\,x^2 - 3\,a\,c^2\,x - \dfrac{b\,c^2}{3} = 0 & (a \neq 0) \quad & (61\,a) \\[3mm] x = -c\,\tan\left(\dfrac{1}{3}\tan^{-1}\left(\dfrac{b}{3\,a\,c}\right)\right) & & (61\,b) \end{cases} \quad (60)$$

By replacing c with $\sqrt{\dfrac{c}{3a}}$, which gives

$$\begin{cases} a\,x^3 + b\,x^2 - c\,x - \dfrac{b\,c}{9\,a} = 0 & (a \neq 0) \quad & (62\,a) \\[3mm] x = -\sqrt{\dfrac{c}{3\,a}}\,\tan\left(\dfrac{1}{3}\tan^{-1}\left(\dfrac{b}{\sqrt{3\,a\,c}}\right)\right) & & (62\,b) \end{cases} \quad (62)$$

Replacing x with x + d gives

$$\begin{cases} a\,x^3 + (b + 3\,a\,d)\,x^2 + \left(3\,a\,d^2 + 2\,b\,d - c\right)x + a\,d^3 + b\,d^2 - c\,d - \dfrac{b\,c}{9\,a} = 0 & (a \neq 0) & (63\,a) \\[3mm] x = -d - \sqrt{\dfrac{c}{3\,a}}\,\tan\left(\dfrac{1}{3}\tan^{-1}\left(\dfrac{b}{\sqrt{3\,a\,c}}\right)\right) & & (63\,b) \end{cases} \quad (63)$$

Replacing b by b – 3ad gives

$$\begin{cases} a x^3 + b x^2 + \left(2 b d - 3 a d^2 - c\right) x + b d^2 - 2 a d^3 - \dfrac{2 c d}{3} - \dfrac{b c}{9 a} = 0 \quad (a \neq 0) & (64\,a) \\[4mm] x = -d - \sqrt{\dfrac{c}{3 a}} \tan\left(\dfrac{1}{3} \tan^{-1}\left(\dfrac{b - 3 a d}{\sqrt{3 a c}}\right)\right) & (64\,b) \end{cases} \quad (64)$$

Replacing c with 2bd–c–3ad² gives

$$\begin{cases} a x^3 + b x^2 + c x + \dfrac{b c}{9 a} + \dfrac{2 c d}{3} - \dfrac{2 b^2 d}{9 a} = 0 \qquad (a \neq 0) & (65\,a) \\[4mm] x = -d - \sqrt{\dfrac{2 b d - c - 3 a d^2}{3 a}} \tan\left(\dfrac{1}{3} \tan^{-1}\left(\dfrac{b - 3 a d}{\sqrt{3 a \left(2 b d - c - 3 a d^2\right)}}\right)\right) & (65\,b) \end{cases} \quad (65)$$

By setting

$$\frac{b c}{9 a} + \frac{2 c d}{3} - \frac{2 b^2 d}{9 a} = D,$$

then multiplying through both sides by 9a, which gives

$$b c + 6 a c d - 2 b^2 d = 9 a D.$$

Solving for d gives

$$d = \frac{9 a D - b c}{6 a c - 2 b^2}.$$

Therefore, we rewrite the expressions,

$$2 b d - c - 3 a d^2 = 2 b \frac{9 a D - b c}{6 a c - 2 b^2} - c - 3 a \left(\frac{9 a D - b c}{6 a c - 2 b^2}\right)^2$$

$$= \frac{4 b \left(9 a D - b c\right)\left(3 a c - b^2\right) - 4 c \left(3 a c - b^2\right)^2 - 3 a \left(9 a D - b c\right)^2}{4 \left(3 a c - b^2\right)^2}$$

$$= \frac{162 a^2 b c D + 9 a b^2 c^2 - 36 a b^3 D - 243 a^3 D^2 - 36 a^2 c^3}{4 \left(3 a c - b^2\right)^2}$$

$$= \frac{9 a \left(18 a b c D + b^2 c^2 - 4 b^3 D - 27 a^2 D^2 - 4 a c^3\right)}{4 \left(3 a c - b^2\right)^2}$$

and

$$b-3ad=b-3a\frac{9aD-bc}{6ac-2b^2}$$

$$=\frac{9abc-2b^3-27a^2D}{2(3ac-b^2)}.$$

Substituting the results of the two expressions above into (65) yields

$$\begin{cases} ax^3+bx^2+cx+D=0 \quad (a\neq 0) & (66a) \\ x=\frac{bc-9aD}{2(3ac-b^2)}-\frac{\sqrt{3(18abcD+b^2c^2-4b^3D-27a^2D^2-4ac^3)}}{2(3ac-b^2)}\tan\left(\frac{1}{3}\tan^{-1}\left(\frac{9abc-2b^3-27a^2D}{3a\sqrt{3(18abcD+b^2c^2-4b^3D-27a^2D^2-4ac^3)}}\right)\right) & (66b) \end{cases} \quad (66)$$

We now can rename D as d to both (66a) and (66b) in (66), and then express (66b) in terms of periodicity of tangent, which gives

$$\begin{cases} ax^3+bx^2+cx+d=0 \quad (a\neq 0) & (67a) \\ x_{k+1}=\frac{bc-9ad}{2(3ac-b^2)} & (67) \\ \qquad -\frac{\sqrt{3(b^2c^2+18abcd-4b^3d-27a^2d^2-4ac^3)}}{2(3ac-b^2)}\tan\left(\frac{1}{3}\tan^{-1}\left(\frac{9abc-2b^3-27a^2d}{3a\sqrt{3(b^2c^2+18abcd-4b^3d-27a^2d^2-4ac^3)}}\right)+\frac{k\pi}{3}\right) & (67b) \\ \text{for } k=0,1,2. \end{cases}$$

Formula (67b) provides solution to the general special equation that cover all real numbers, since the domain of inverse tangent is defined in the interval $(-\infty, \infty)$. If the sign of $(b^2c^2+18abcd-4b^3d-27a^2d^2-4ac^3)$ is negative, it leads to the appearance of an imaginary unit i. In such case, the solution to the equation is in the form of $\tanh\left(\frac{1}{3}\tanh^{-1}[\ ]\right)$ for the domain defined in $(-1, 1)$, and in the form of $\coth\left(\frac{1}{3}\coth^{-1}[\ ]\right)$ for the domain defined in $(-\infty, -1)$ or $(1, \infty)$. We will elaborate on these points in the following Special Cases section, where we will provide further proofs. In addition, we observe that

$$27a^2(b^2c^2+18abcd-4b^3d-4ac^3-27a^2d^2)=4(b^2-3ac)^3-(27a^2d+2b^3-9abc)^2$$

Then formula (67b) can be rewritten as

$$x_{k+1}=\frac{9ad-bc}{2(b^2-3ac)}+\frac{\sqrt{4(b^2-3ac)^3-(9abc-27a^2d-2b^3)^2}}{6a(b^2-3ac)}\tan\left(\frac{1}{3}\tan^{-1}\left(\frac{9abc-27a^2d-2b^3}{\sqrt{4(b^2-3ac)^3-(27a^2d+2b^3-9abc)^2}}\right)+\frac{\pi k}{3}\right). \quad (67c)$$

Special Cases in (67b)

1. Case $(b^2c^2+18abcd-4b^3d-27a^2d^2-4ac^3) = 0$:

In this case, formula (67b) provides three simple roots, two of which are identical:

$$x_{1,3} = \frac{9\,a\,d - b\,c}{2\left(b^2 - 3\,a\,c\right)} \quad \text{for k = 0, 2}$$

and

$$x_2 = \frac{4\,a\,b\,c - b^3 - 9\,a^2\,d}{a\left(b^2 - 3\,a\,c\right)} \quad \text{for k = 1.}$$

Proof

- At k = 0, we observe that $\tan^{-1}(1/0)$ approaches $\pi/2$ By substitution, we obtain

$$x_1 = \frac{9\,a\,d - b\,c}{2\left(b^2 - 3\,a\,c\right)} + \frac{0}{2\left(b^2 - 3\,a\,c\right)}\tan\left(\frac{\pi}{6}\right)$$

$$= \frac{9\,a\,d - b\,c}{2\left(b^2 - 3\,a\,c\right)}.$$

- Similarly, at k = 2, we get

$$x_3 = \frac{9\,a\,d - b\,c}{2\left(b^2 - 3\,a\,c\right)} + \frac{0}{2\left(b^2 - 3\,a\,c\right)}\tan\left(\frac{\pi}{6} + \frac{2\pi}{3}\right)$$

$$= \frac{9\,a\,d - b\,c}{2\left(b^2 - 3\,a\,c\right)}.$$

- At k = 1, we have

$$x_2 = \frac{9\,a\,d - b\,c}{2\left(b^2 - 3\,a\,c\right)} + \frac{0}{2\left(b^2 - 3\,a\,c\right)}\tan\left(\frac{\pi}{6} + \frac{\pi}{3}\right)$$

$$= \frac{9\,a\,d - b\,c}{2\left(b^2 - 3\,a\,c\right)} + 0*\infty, \quad \text{which is an indeterminate form.}$$

To evaluate whether the limit of x_2 exists, we set

$$\sqrt{3\left(3\,b^2\,c^2 + 54\,a\,b\,c\,d - 12\,b^3\,d - 12\,a\,c^3 - 81\,a^2\,d^2\right)} = h$$

and

$$\frac{9\,a\,b\,c - 27\,a^2\,d - 2\,b^3}{3\,a} = A.$$

Then x_2 from formula (67b) is rewritten in terms of h and A as indicated in the following limit:

$$x_2 = \lim_{h \to 0} \left\{ \frac{9\,a\,d - b\,c}{2\left(b^2 - 3\,a\,c\right)} + \frac{1}{2\left(b^2 - 3\,a\,c\right)} h \tan\left(\frac{1}{3} \tan^{-1}\left(\frac{A}{h}\right) + \frac{\pi}{3}\right)\right\}$$

As h tends to 0, it results in (0)(tan(π/2)) or gives an indeterminate form of (0)*(∞) in which the value ∞ is only relevant to cosine. Therefore, the above expression is simplified and rewritten as

$$x_2 = \lim_{h \to 0} \left\{ \frac{9\,a\,d - b\,c}{2\left(b^2 - 3\,a\,c\right)} + \frac{1}{2\left(b^2 - 3\,a\,c\right)} \frac{h}{\cos\left(\frac{1}{3} \tan^{-1}\left(\frac{A}{h}\right) + \frac{\pi}{3}\right)}\right\}.$$

Applying L'Hôpital's rule to h gives

$$= \lim_{h \to 0} \left\{ \frac{9\,a\,d - b\,c}{2\left(b^2 - 3\,a\,c\right)} + \frac{1}{2\left(b^2 - 3\,a\,c\right)} \frac{1}{\dfrac{A \sin\left(\frac{1}{3} \tan^{-1}\left(\frac{A}{h}\right) + \frac{\pi}{3}\right)}{3\left(h^2 + A^2\right)}}\right\}$$

$$= \frac{9\,a\,d - b\,c}{2\left(b^2 - 3\,a\,c\right)} + \frac{1}{2\left(b^2 - 3\,a\,c\right)} 3A$$

$$= \frac{9\,a\,d - b\,c}{2\left(b^2 - 3\,a\,c\right)} + \frac{9\,a\,b\,c - 27\,a^2\,d - 2\,b^3}{2\,a\left(b^2 - 3\,a\,c\right)}$$

$$= \frac{4\,a\,b\,c - b^3 - 9\,a^2\,d}{a\left(b^2 - 3\,a\,c\right)},$$

which we complete the proof.

Example. *Find the roots of the cubic equation* $3x^3 - 16x^2 + 28x - 16 = 0$.

The coefficients of the given equation are a = 3, b = -16, c = 28, and d = -16. When we substitute these values into solution formula (67b) or (67c), it fails because the denominator is zero, namely

$$b^2 c^2 + 18\,a\,b\,c\,d - 4\,b^3 d - 4\,a\,c^3 - 27\,a^2\,d^2 =$$

$$(-16)^2 (28)^2 + 18(3)(-16)(28)(-16) - 4(-16)^3(-16) - 27\left(3^2\right)(-16)^2 - 4(3)(28)^3 = 0.$$

In this case, the formula from Special Case previously described is used to provide the solution to given cubic equation as follows:

$$x_{1,3} = \frac{9ad-bc}{2(b^2-3ac)} = \frac{9(3)(-16)-(-16)(28)}{2((-16)^2-3(3)(28))} = 2$$

$$x_2 = \frac{4abc-b^3-9a^2d}{a(b^2-3ac)} = \frac{4(3)(-16)(28)-(-16)^3-9(3^2)(-16)}{3((-16)^2-3(3)(28))} = \frac{4}{3}$$

2. Case $(b^2c^2+18abcd-4b^3d-27a^2d^2-4ac^3) < 0$:

i. If $-1 < \dfrac{9abc-27a^2d-2b^3}{3a\sqrt{3(|b^2c^2+18abcd-4b^3d-4ac^3-27a^2d^2|)}} < 1$:

By applying identities (K) and (L) to (67c), which can be rewritten as

$$x = \frac{9ad-bc}{2(b^2-3ac)} + \frac{\sqrt{3(|b^2c^2+18abcd-4b^3d-4ac^3-27a^2d^2|)}}{2(b^2-3ac)}\tanh\left(\frac{1}{3}\tanh^{-1}\left(\frac{9abc-27a^2d-2b^3}{3a\sqrt{3(|b^2c^2+18abcd-4b^3d-4ac^3-27a^2d^2|)}}\right)+\frac{i\pi k}{3}\right)$$

$$(67d)$$

ii. If $\left|\dfrac{9abc-27a^2d-2b^3}{3a\sqrt{3(b^2c^2+18abcd-4b^3d-4ac^3-27a^2d^2)}}\right| > 1$:

We apply the identity,

$$\tanh^{-1}u = \frac{i\pi}{2} + \coth^{-1}u,$$

to the following expression:

$$\tanh\left(\frac{1}{3}\tanh^{-1}(u)+\frac{i\pi k}{3}\right) = \tanh\left(\frac{i\pi}{6}+\frac{1}{3}\coth^{-1}(u)+\frac{i\pi k}{3}\right)$$

$$= \tanh\left(\frac{i\pi}{2}-\frac{i\pi}{3}+\frac{1}{3}\tanh^{-1}(u)+\frac{i\pi k}{3}\right)$$

$$= \tanh\left(\frac{i\pi}{2}+\frac{1}{3}\tanh^{-1}(u)+\frac{i\pi(k-1)}{3}\right)$$

$$= \coth\left(\frac{1}{3}\coth^{-1}(u)+\frac{i\pi(k-1)}{3}\right).$$

By re-indexing the value of k with k + 1, therefore, formula (67d) can be rewritten as

$$x_{k+1} = \frac{9ad-bc}{2(b^2-3ac)} + \frac{\sqrt{3(|b^2c^2+18abcd-4b^3d-4ac^3-27a^2d^2|)}}{2(b^2-3ac)}\coth\left(\frac{1}{3}\coth^{-1}\left(\frac{9abc-27a^2d-2b^3}{3a\sqrt{3(|b^2c^2+18abcd-4b^3d-4ac^3-27a^2d^2|)}}\right)+\frac{i\pi k}{3}\right)$$

3. Case b²–3ac = 0:

If b²–3ac = 0 or $b=\pm\sqrt{3ac}$, formula (67b) or (67c) has a form of 0/0. In this case, to derive the cubic roots from (67b) or (67c), we can take the limit of b as it approaches $\pm\sqrt{3ac}$, which results in a similar formula to (43) or (44), namely

$$x_{k+1}=-\frac{\pm\sqrt{3ac}}{3a}+\frac{1}{3a}\sqrt[3]{\pm(3ac)^{3/2}-27a^2d}\,e^{(2i\pi k/3)},\ k=0,1,2. \qquad (67f)$$

The value of ± is determined by the sign of b. The positive sign is used if b ≥ 0; otherwise, the negative sign is used. Alternatively, to eliminate the need for the ± symbol, formula (67f) can be expressed involving b, resulting in the following expression:

$$x_{k+1}=-\frac{b}{3a}+\frac{1}{3a}\sqrt[3]{b^3-27a^2d}\,e^{(2i\pi k/3)},\ k=0,1,2. \qquad (67g)$$

4. Case b²–3ac = 0, 9ad–bc = 0 and 9abc–27a²d–2b³ = 0:

In this case, the equation has a repeated root. Indeed, by applying L'Hôpital's rule on (67c) for the following quotients, which give

i. $\displaystyle\lim_{b\to\pm\sqrt{3ac}}\frac{9ad-bc}{2(b^2-3ac)}=-\frac{c}{4b}$

ii. $\displaystyle\lim_{b\to\pm\sqrt{3ac}}\frac{\sqrt{4(b^2-3ac)^3-(27a^2d+2b^3-9abc)^2}}{6a(b^2-3ac)}=\lim_{b\to\pm\sqrt{3ac}}\frac{i(27a^2d+2b^3-9abc)}{6a(b^2-3ac)}$

$\displaystyle\qquad\qquad =\lim_{b\to\pm\sqrt{3ac}}\frac{i(6b^2-9ac)}{12ab}$

$\displaystyle\qquad\qquad =\frac{i(2b^2-3ac)}{4ab}$ (preferable form)

$\displaystyle\qquad\qquad =\frac{i\sqrt{3ac}}{\pm4a}$ (less preferred form)

Notice that we avoid using the last form due to potential ambiguity resulting from the plus or minus sign. Instead, preference is given to the preceding form.

iii. $\displaystyle\lim_{b\to\pm\sqrt{3ac}}\tan\left(\frac{1}{3}\tan^{-1}\left(\frac{9abc-27a^2d-2b^3}{\sqrt{4(b^2-3ac)^3-(27a^2d+2b^3-9abc)^2}}\right)\right)=i\tanh\left(\frac{1}{3}\tanh^{-1}(i)\right).$

Therefore, by putting all expressions together, formula (67c) yields

$$x_{k+1}=-\frac{c}{4b}-\frac{2b^2-3ac}{4ab}\tanh\left(\frac{1}{3}\tanh^{-1}(1)+\frac{\pi ik}{3}\right), k=0,1,2. \tag{67h}$$

But we have $\tanh\left(\frac{1}{3}\tanh^{-1}(1)+\frac{\pi ik}{3}\right)=1$ [see the proof in Appendix 2-A-10], which (67h) can be deduced to yield a repeated root, namely

$$x_1=x_2=x_3=\frac{ac-b^2}{2ab}. \tag{67i}$$

Example. Find the roots of the equation $8x^3-36x^2+54x-27 = 0$.

Solution. The coefficients of the given equation are a = 8, b = -36, c = 54, and d = -27. Formula (67b) or (67c) cannot be applied because all of the following quotients become indeterminate if plugging in the coefficients, namely

- $$\frac{9ad-bc}{2(b^2-3ac)}=-\frac{9(8)(-27)-(-36)(54)}{4(-36)}=\frac{0}{0}$$

- $$\frac{\sqrt{4(b^2-3ac)^3-(27a^2d+2b^3-9abc)^2}}{6a(b^2-3ac)}=\frac{\sqrt{4((-36)^2-3(8)(54))^3-(27(8^2)(-27)+2(-36)^3-9(8)(-36)(-27))^2}}{6(8)((-36)^2-3(8)(54))}=\frac{0}{0}$$

- $$\frac{9abc-27a^2d-2b^3}{\sqrt{4(b^2-3ac)^3-(27a^2d+2b^3-9abc)^2}}\frac{9(8)(-36)(54)-27(8^2)(-27)-2(-36)^3}{\sqrt{4((-36)^2-3(8)(54))^3-(27(8)^2(-27)+2(-36)^3-9(8)(-36)(54))^2}}=\frac{0}{0}$$

Applying formula (67e) gives

$$x_1=x_2=x_3=\frac{ac-b^2}{2ab}=\frac{8(54)-(-36)^2}{2(8)(-36)}=\frac{3}{2}.$$

Summary of Solution to Cubic Equation (Type 3)

Cubic equation: $ax^3+bx^2+cx+d=0 \ (a\neq0)$

Define

$$D=3(b^2c^2+18abcd-4b^3d-4ac^3-27a^2d^2) \tag{67k}$$

and

$$T = \frac{9\,abc - 27\,a^2 d - 2\,b^3}{3\,a\sqrt{D}}$$

(67 *l*)

1. Case D > 0 and b²−3ac ≠ 0:

The equation has three distinct real roots, given by formula (67b) or (67c):

$$x_{k+1} = \frac{9\,ad - bc}{2(b^2 - 3\,ac)} + \frac{\sqrt{D}}{2(b^2 - 3\,ac)} \tan\left(\frac{1}{3}\tan^{-1}(T) + \frac{\pi k}{3}\right), k = 0,1,2,$$

(67 *m*)

By applying (46g), we convert formula (67m) to the radical form,

$$x_{k+1} = \frac{9\,ad - bc}{2(b^2 - 3\,ac)} + \frac{\sqrt{D}}{i\,2(b^2 - 3\,ac)} \frac{(1+iT)^{1/3}e^{2i\pi k/3} - (1-iT)^{1/3}}{(1+iT)^{1/3}e^{2i\pi k/3} + (1-iT)^{1/3}}, k = 0,1,2.$$

(67 *n*)

2. If D = 0 (or T is undefined):

The equation has three simple roots, two of which are identical:

$$x_{1,3} = \frac{9\,ad - bc}{2(b^2 - 3\,ac)}$$

(67 *o*)

and

$$x_2 = \frac{4\,abc - b^3 - 9\,a^2 d}{a(b^2 - 3\,ac)}.$$

(67 *p*)

(See Special Case 1.)

3. Case D<0:

T is rewritten as

$$T = \frac{9\,abc - 27\,a^2 d - 2\,b^3}{3\,a\sqrt{|D|}}.$$

(67 *q*)

Considering two possible scenarios:

a. a. If -1<T<1, the equation has one real root and two complex conjugate roots, provided by formula (67d):

$$x_{k+1} = \frac{9\,ad - bc}{2(b^2 - 3\,ac)} + \frac{\sqrt{|D|}}{2(b^2 - 3\,ac)} \tanh\left(\frac{1}{3}\tanh^{-1}(T) + \frac{i\pi k}{3}\right), k = 0,1,2,$$

(67 *r*)

112

By applying (46c), formula (67r) is transformed into the radical form,

$$x_{k+1}=\frac{9ad-bc}{2(b^2-3ac)}+\frac{\sqrt{|D|}}{2(b^2-3ac)}\frac{(1+T)^{1/3}e^{2i\pi k/3}-(1-T)^{1/3}}{(1+T)^{1/3}e^{2i\pi k/3}+(1-T)^{1/3}},\quad k=0,1,2. \qquad (67s)$$

b. If T < -1 or T > 1, the equation has one root and two complex conjugate roots via formula (67e):

$$x_{k+1}=\frac{9ad-bc}{2(b^2-3ac)}+\frac{\sqrt{|D|}}{2(b^2-3ac)}\coth\left(\frac{1}{3}\coth^{-1}(T)+\frac{i\pi k}{3}\right),k=0,1,2. \qquad (67t)$$

(See Special Case 2.)

By applying (46d), we convert formula (67t) to the radical form,

$$x_{k+1}=\frac{9ad-bc}{2(b^2-3ac)}+\frac{\sqrt{|D|}}{2(b^2-3ac)}\frac{(T+1)^{1/3}e^{2i\pi k/3}+(T-1)^{1/3}}{(T+1)^{1/3}e^{2i\pi k/3}-(T-1)^{1/3}},\quad k=0,1,2. \qquad (67u)$$

4. Case 3ac−b² = 0:
the equation has one real root and two complex conjugate roots, given by

$$x_{k+1}=-\frac{\pm\sqrt{3ac}}{3a}+\frac{1}{3a}\sqrt[3]{\pm(3ac)^{3/2}-27a^2d\,e^{(2i\pi k/3)}},\quad k=0,1,2. \qquad (67v)$$

The choice of ± depends on the sign of b. The positive sign is used if b ≥ 0; otherwise, the negative sign is used. The formula can also be expressed in terms of b,

$$x_{k+1}=-\frac{b}{3a}+\frac{1}{3a}\sqrt[3]{b^3-27a^2d\,e^{(2i\pi k/3)}},\quad k=0,1,2. \qquad (67x)$$

(See Special Case 3.)

5. If b²−3ac = 0, 9ad−bc = 0 and 9abc−27a²d−2b³ = 0,
the equation has three simple roots and they are identical:

$$x_1=x_2=x_3=\frac{ac-b^2}{2ab}. \qquad (67z)$$

(See Special Case 4.)

Note:

- i is the imaginary unit.

- $e^{i2\pi/3} = \frac{1}{2}\left(-1+i\sqrt{3}\right)$ and $e^{i4\pi/3} = \frac{1}{2}\left(-1-i\sqrt{3}\right)$.

Example. Use cubic formula (type 3) to find the roots of the cubic equation,

$$x^3 + x^2 + 3x + 5 = 0.$$

Solution. To find the cubic roots of the given equation, follow these steps:
1. Identify the coefficients: a = 1, b = 1, c = 3, and d = 5.
2. Determine the values of D and T:

$$D = 3\left(b^2 c^2 + 18\, a\, b\, c\, d - 4\, b^3 d - 4\, a\, c^3 - 27\, a^2 d^2\right)$$
$$= 3\left(1^2 3^2 + 18 \cdot 1 \cdot 1 \cdot 3 \cdot 5 - 4 \cdot 1^3 \cdot 5 - 4 \cdot 1 \cdot 3^2 - 27 \cdot 1^2 \cdot 5^2\right)$$
$$= -1572 < 0.$$

Because D < 0, it follows that (67q) is used to determine T:

$$T = \frac{9\, a\, b\, c - 27\, a^2 d - 2 b^3}{3\, a\sqrt{|D|}}$$
$$= \frac{9 \cdot 1 \cdot 1 \cdot 3 - 27 \cdot 1^2 \cdot 5 - 2 \cdot 1^3}{3 \cdot 1 \cdot \sqrt{-1572}}$$
$$= -\frac{55}{3\sqrt{393}}$$
$$= -0.9247\ldots$$

Since D < 0 and -1<T<1, formula (67r) is applied. By substituting the coefficients, we obtain the roots,

$$x_1 = -\frac{21}{8} - \frac{1}{8}\sqrt{393}\,\tanh\left(\frac{1}{3}\tanh^{-1}\left(-\frac{55}{3\sqrt{393}}\right)\right),$$

$$x_2 = -\frac{21}{8} - \frac{1}{8}\sqrt{393}\,\tanh\left(\frac{1}{3}\tanh^{-1}\left(-\frac{55}{3\sqrt{393}}\right) + \frac{i2\pi}{3}\right)$$

and

$$x_3 = -\frac{21}{8} - \frac{1}{8}\sqrt{393}\,\tanh\left(\frac{1}{3}\tanh^{-1}\left(-\frac{55}{3\sqrt{393}}\right) + \frac{i4\pi}{3}\right).$$

Additionally, applying formula (67s) gives the solution in radical form as follows:

$$x_1 = -\frac{21}{8} - \frac{\sqrt{393}}{8} \frac{\left(\dfrac{3\sqrt{393}-55}{3\sqrt{393}}\right)^{1/3} + \left(\dfrac{-3\sqrt{393}-55}{3\sqrt{393}}\right)^{1/3}}{\left(\dfrac{3\sqrt{393}-55}{3\sqrt{393}}\right)^{1/3} - \left(\dfrac{-3\sqrt{393}-55}{3\sqrt{393}}\right)^{1/3}}$$

$$= -\frac{21}{8} - \frac{\sqrt{393}}{8} \frac{\left(3\sqrt{393}-55\right)^{1/3} - \left(3\sqrt{393}+55\right)^{1/3}}{\left(3\sqrt{393}-55\right)^{1/3} + \left(3\sqrt{393}+55\right)^{1/3}},$$

$$x_2 = -\frac{21}{8} + \frac{\sqrt{393}}{8} \frac{\left(3\sqrt{393}-55\right)^{1/3}\left(-1+i\sqrt{3}\right) - 2\left(3\sqrt{393}+55\right)^{1/3}}{\left(3\sqrt{393}-55\right)^{1/3}\left(-1+i\sqrt{3}\right) + 2\left(3\sqrt{393}+55\right)^{1/3}}$$

and

$$x_3 = -\frac{21}{8} + \frac{\sqrt{393}}{8} \frac{\left(3\sqrt{393}-55\right)^{1/3}\left(-1-i\sqrt{3}\right) - 2\left(3\sqrt{393}+55\right)^{1/3}}{\left(3\sqrt{393}-55\right)^{1/3}\left(-1-i\sqrt{3}\right) + 2\left(3\sqrt{393}+55\right)^{1/3}}.$$

Section III. Solving General Quartic Equation

The general quartic equation [9] is expressed as

$$a x^4 + b x^3 + c x^2 + d x + e = 0 \quad (a \neq 0), \tag{70}$$

where a, b, c, d and e are real numbers.

We use (20) as our starting journey to seek solution to the quartic equation in the following steps:

1. Define

$$T_k = \left(1 - \cos\left[\frac{1}{2}\cos^{-1}\left(1 - \frac{8ac}{b^2}\right) + \pi k\right]\right), \quad k = 0,1.$$

2. Rewriting (20) in terms of T_k gives

$$\begin{cases} a x^2 + b x + c = 0 \quad (a \neq 0) \\ x_{k+1} = -\dfrac{b}{2a} T_k \end{cases} \tag{71}$$

3. By replacing x_{k+1} with $\frac{x}{d}+\frac{e}{x}$ into (71), expanding the terms and simplifying, which gives

$$
\begin{cases}
a x^4 + b d x^3 + \left(2 a d e + c d^2\right) x^2 + b d^2 e x + a d^2 e^2 = 0 \quad (a \neq 0) & (72\,a) \\
x^2 + \dfrac{b T_k}{2 a} x + d e = 0 & (72\,b)
\end{cases} \tag{72}
$$

4. Solving for x from quadratic equation (72b) gives

$$
\begin{cases}
a x^4 + b d x^3 + \left(2 a d e + c d^2\right) x^2 + b d^2 e x + a d^2 e^2 = 0 \quad (a \neq 0) & (73\,a) \\
x_{j+1} = -\dfrac{b d\, T_k}{4 a}\left(1 - \cos\left[\dfrac{1}{2}\cos^{-1}\left(1 - \dfrac{32\, a^2 e}{b^2 d\, T^2_{\,k}}\right) + \pi\, j\right]\right),\ j = 0, 1. & (73\,b)
\end{cases} \tag{73}
$$

For each value of j, formula (73b) gives two roots, corresponding to k = 0 and k = 1. Therefore, it gives a total of four roots for equation (73a), corresponding to the combination of j = 0 and j = 1, as follows:

$$
\begin{cases}
x_1 = -\dfrac{b d\, T_0}{4 a}\left(1 - \cos\left[\dfrac{1}{2}\cos^{-1}\left(1 - \dfrac{32\, a^2 e}{b^2 d\, T^2_{\,0}}\right)\right]\right) & (73\,c) \\[4mm]
x_2 = -\dfrac{b d\, T_1}{4 a}\left(1 - \cos\left[\dfrac{1}{2}\cos^{-1}\left(1 - \dfrac{32\, a^2 e}{b^2 d\, T^2_{\,1}}\right)\right]\right) & (73\,d)
\end{cases}
$$

$$
\begin{cases}
x_3 = -\dfrac{b d\, T_0}{4 a}\left(1 - \cos\left[\dfrac{1}{2}\cos^{-1}\left(1 - \dfrac{32\, a^2 e}{b^2 d\, T^2_{\,0}}\right)\right]\right) & (73\,e) \\[4mm]
x_4 = -\dfrac{b d\, T_1}{4 a}\left(1 - \cos\left[\dfrac{1}{2}\cos^{-1}\left(1 - \dfrac{32\, a^2 e}{b^2 d\, T^2_{\,1}}\right)\right]\right) & (73\,f)
\end{cases}
$$

To ensure consistency in trigonometric function usage, we refrain from referencing Appendix [2-A-5] to simplify formula (73b).

5. By replacing b with b/d into (73) and rearranging the terms, which gives

$$\begin{cases} a\,x^4+b\,x^3+(2\,a\,d\,e+c\,d^2)\,x^2+b\,d\,e\,x+a\,d^2\,e^2=0 \quad (a\neq0) & (74\,a) \\ x_{j+1}=-\dfrac{b\,T_k}{4\,a}\left(1-\cos\left[\dfrac{1}{2}\cos^{-1}\left(1-\dfrac{32\,a^2\,d\,e}{b^2\,T_k^2}\right)+j\,\pi\right]\right),\ j=0,1. & (74\,b) \end{cases} \quad (74)$$

where

$$T_k=\left(1-\cos\left[\dfrac{1}{2}\cos^{-1}\left(1-\dfrac{8\,a\,c\,d^2}{b^2}\right)+\pi\,k\right]\right),\quad k=0,1.$$

6. By replacing c with $(c-2ade)/d^2$ into (74) and rearranging the terms, which gives

$$\begin{cases} a\,x^4+b\,x^3+c\,x^2+b\,d\,e\,x+a\,d^2\,e^2=0 \quad (a\neq0) & (75\,a) \\ x_{j+1}=-\dfrac{b\,T_k}{4\,a}\left(1-\cos\left[\dfrac{1}{2}\cos^{-1}\left(1-\dfrac{32\,a^2\,d\,e}{b^2\,T_k^2}\right)+\pi\,j\right]\right),\ j=0,\,1. & (75\,b) \end{cases} \quad (75)$$

where

$$T_k=\left(1-\cos\left[\dfrac{1}{2}\cos^{-1}\left(1-\dfrac{8\,a\,(c-2\,a\,d\,e)}{b^2}\right)+\pi\,k\right]\right),\,k=0,1.$$

7. Replacing d with $d/(be)$ and simplifying further gives

$$\begin{cases} a\,x^4+b\,x^3+c\,x^2+d\,x+\dfrac{a\,d^2}{b^2}=0 \quad (a\neq0) & (76\,a) \\ x_{j+1}=-\dfrac{b\,T_k}{4\,a}\left(1-\cos\left[\dfrac{1}{2}\cos^{-1}\left(1-\dfrac{32\,a^2\,d}{b^3\,T_k^2}\right)+\pi\,j\right]\right),\ j=0,\,1, & (76\,b) \end{cases} \quad (76)$$

where $T_k=\left(1-\cos\left[\dfrac{1}{2}\cos^{-1}\left(1-\dfrac{8\,a\,(b\,c-2\,a\,d)}{b^3}\right)+\pi\,k\right]\right),\,k=0,1.$

Equation (76a) is an incomplete quartic equation lacking the coefficient e. In the next four steps, we illustrate how to construct this coefficient e, thereby completing the quartic equation to its general form.

8. Let p be a real number. Replacing x with $(x+p)$ gives

$$\begin{cases} ax^4+(b+4a\rho)x^3+(c+6a\rho^2+3b\rho)x^2+(d+4a\rho^3+3b\rho^2+2c\rho)x+\dfrac{ad^2}{b^2}+a\rho^4+b\rho^3+c\rho^2+d\rho=0 \ (a\neq0) & (77a) \\[4mm] x_{j+1}=-\rho-\dfrac{bT_k}{4a}\left(1-\cos\left[\dfrac{1}{2}\cos^{-1}\left(1-\dfrac{32a^2d}{b^3T_k^2}\right)+\pi j\right]\right), j=0,1, & (77b) \end{cases} \quad (77)$$

where T_k is the same in the step 6.

9. Replacing b with b-4aρ and (b-4aρ) ≠ 0 gives

$$\begin{cases} ax^4+bx^3+(c-6a\rho^2+3b\rho)x^2+(d-8a\rho^3+3b\rho^2+2c\rho)x+\dfrac{ad^2}{(b-4a\rho)^2}-3a\rho^4+b\rho^3+c\rho^2+d\rho=0 \ (a\neq0) & (78a) \\[4mm] x_{j+1}=-\rho-\dfrac{(b-4a\rho)T_k}{4a}\left(1-\cos\left[\dfrac{1}{2}\cos^{-1}\left(1-\dfrac{32a^2d}{(b-4a\rho)^3T_k^2}\right)+\pi j\right]\right) j=0,1, & (78b) \end{cases} \quad (78)$$

where $T_k=\left(1-\cos\left[\dfrac{1}{2}\cos^{-1}\left(1-\dfrac{8a(bc-4ac\rho-2ad)}{(b-4a\rho)^3}\right)+\pi k\right]\right), k=0,1.$

10. Replacing c with (c+6aρ²-3bρ) gives

$$\begin{cases} ax^4+bx^3+cx^2+(d+4a\rho^3-3b\rho^2+2c\rho)x+3a\rho^4-2b\rho^3+c\rho^2+d\rho+\dfrac{ad^2}{(b-4a\rho)^2}=0 \ (a\neq0) & (79a) \\[4mm] x_{j+1}=-\rho-\dfrac{(b-4a\rho)T_k}{4a}\left(1-\cos\left[\dfrac{1}{2}\cos^{-1}\left(1-\dfrac{32a^2d}{(b-4a\rho)^3T_k^2}\right)+\pi j\right]\right), j=0,1, & (79b) \end{cases} \quad (79)$$

where

$$T_k=1-\cos\left[\dfrac{1}{2}\cos^{-1}\left(1+\dfrac{8a(24a^2\rho^3-18ab\rho^2+(4ac+3b^2)\rho+2ad-bc)}{(b-4a\rho)^3}\right)+\pi k\right], \ k=0,1.$$

11. Replacing d with (d + 3bρ² - 4aρ³ - 2cρ) gives

$$ax^4+bx^3+cx^2+dx+\frac{\left(b^3+8a^2d-4abc\right)\rho^3+\left(4ac^2-2abd-b^2c\right)\rho^2+\left(b^2d-4acd\right)\rho+ad^2}{\left(b-4a\rho\right)^2}=0 \ \left(a\neq0\right) \quad (80a)$$

$$x_{j+1}=-\rho-\frac{\left(b-4ap\right)T_k}{4a}\left(1-\cos\left[\frac{1}{2}\cos^{-1}\left(1-\frac{32a^2\left(d+3b\rho^2-4a\rho^3-2c\rho\right)}{\left(b-4a\rho\right)^3T_k^2}\right)+\pi j\right]\right),j=0,1, \ (60b) \qquad (80)$$

where

$$T_k=1-\cos\left[\frac{1}{2}\cos^{-1}\left(1-\frac{8a\left(12ab\rho^2+bc-16a^2\rho^3-3b^2\rho-2ad\right)}{\left(b-4a\rho\right)^3}\right)+\pi k\right], \ k=0,1. \qquad (80c)$$

We note that $\left(b-4a\rho\right)^3-4a\left(12ab\rho^2-16a^2\rho^3-3b^2\rho+bc-2ad\right)=b^3+8a^2d-4abc.$

Hence, we rewrite the expression,

$$1-\frac{8a\left(12ab\rho^2-16a^2\rho^3-3b^2\rho+bc-2ad\right)}{\left(b-4a\rho\right)^3}=2-\frac{8a\left(12ab\rho^2-16a^2\rho^3-3b^2\rho+bc-2ad\right)}{\left(b-4a\rho\right)^3}-1$$

$$=\frac{2\left(b^3+8a^2d-4abc\right)}{\left(b-4a\rho\right)^3}-1.$$

12. Setting

$$\frac{\left(b^3+8a^2d-4abc\right)\rho^3+\left(4ac^2-2abd-b^2c\right)\rho^2+\left(b^2d-4acd\right)\rho+ad^2}{\left(b-4a\rho\right)^2}=e. \qquad (81)$$

Multiplying both sides of (81) by $(b-4a\rho)^2$ and moving all terms to the left-hand side of the equal sign gives the cubic equation in ρ:

$$\left(b^3+8a^2d-4abc\right)\rho^3+\left(4ac^2-2abd-b^2c-16a^2e\right)\rho^2+\left(b^2d+8abe-4acd\right)\rho+ad^2-b^2e=0. \qquad (82)$$

By using cubic formula (40b) to solve for ρ, which gives:

$$\rho=\frac{16a^2e+b^2c+2abd-4ac^2}{3\left(b^3+8a^2d-4abc\right)}+\frac{2\sqrt{\left(4ac^2-2abd-b^2c-16a^2e\right)^2-3\left(b^3+8a^2d-4abc\right)\left(b^2d+8abe-4acd\right)}}{3\left(b^3+8a^2d-4abc\right)}*\cos\left(\frac{1}{3}\cos^{-1}(V)\right) \qquad (83)$$

where

$$V=\frac{9\left(b^3+8a^2d-4abc\right)\left(4ac^2-2abd-b^2c-16a^2e\right)\left(b^2d+8abe-4acd\right)-2\left(4ac^2-2abd-b^2c-16a^2e\right)^3-27\left(b^3+8a^2d-4abc\right)^2\left(ad^2-b^2e\right)}{2\left(\left(4ac^2-2abd-b^2c-16a^2e\right)^2-3\left(b^3+8a^2d-4abc\right)\left(b^2d+8abe-4acd\right)\right)^{3/2}}$$

Solution to the Quartic Formula

We obtain the solution to the quartic equation by substituting the expression involving e as indicated in (81) into (80a) and value of ρ from (83) into (80b), resulting in:

$$ax^4+bx^3+cx^2+dx+e=0 \quad (a\neq 0) \tag{84a}$$

$$x_{j+1}=-\rho-\frac{(b-4a\rho)T_k}{4a}\left(1-\cos\left[\frac{1}{2}\cos^{-1}\left(1-\frac{32a^2(d+3b\rho^2-4a\rho^3-2c\rho)}{(b-4a\rho)^3T_k^2}\right)+\pi j\right]\right),j=0,1. \tag{84b}$$

where $\hspace{8cm}$ (84)

$$T_k=1-\cos\left[\frac{1}{2}\cos^{-1}\left(\frac{2(b^3+8a^2d-4abc)}{(b-4a\rho)^3}-1\right)+\pi k\right], k=0,1. \tag{84c}$$

$$\rho=\frac{16a^2e+b^2c+2abd-4ac^2}{3(b^3+8a^2d-4abc)}+\frac{2\sqrt{(4ac^2-2abd-b^2c-16a^2e)^2-3(b^3+8a^2d-4abc)(b^2d+8abe-4acd)}}{3(b^3+8a^2d-4abc)}*\cos\left(\frac{1}{3}\cos^{-1}(V)\right) \tag{84d}$$

$$V=\frac{9(b^3+8a^2d-4abc)(4ac^2-2abd-b^2c-16a^2e)(b^2d+8abe-4acd)-2(4ac^2-2abd-b^2c-16a^2e)^3-27(b^3+8a^2d-4abc)^2(ad^2-b^2e)}{2((4ac^2-2abd-b^2c-16a^2e)^2-3(b^3+8a^2d-4abc)(b^2d+8abe-4acd))^{3/2}}$$

We expect the value of ρ derived from cubic formula to be a real root under assumption that all coefficients are real. It is stated not just as $\cos((1/3)\cos^{-1}(V))$ for $-1\leq V\leq 1$, but also as $\cosh((1/3)\cosh^{-1}(V))$ for $V > 1$ or $V < -1$ provided that the cubic equation delivers a real root. More information can be found in the Solution to Cubic Equation section. Formula (84b) provides four roots to quartic equation (84a). The first two roots corresponding to j = 0 for both k = 0 and k = 1 are:

$$x_1=-\rho-\frac{(b-4a\rho)T_0}{4a}\left(1-\cos\left[\frac{1}{2}\cos^{-1}\left(1-\frac{32a^2(d-4a\rho^3+3b\rho^2-2c\rho)}{(b-4a\rho)^3T_0^2}\right)\right]\right) \tag{84e}$$

$$=-\rho-\frac{(b-4a\rho)T_0}{2a}\left(\sin^2\left[\frac{1}{4}\cos^{-1}\left(1-\frac{32a^2(d-4a\rho^3+3b\rho^2-2c\rho)}{(b-4a\rho)^3T_0^2}\right)\right]\right) \tag{84f}$$

$$x_2=-\rho-\frac{(b-4a\rho)T_1}{4a}\left(1-\cos\left[\frac{1}{2}\cos^{-1}\left(1-\frac{32a^2(d-4a\rho^3+3b\rho^2-2c\rho)}{(b-4a\rho)^3T_1^2}\right)\right]\right) \tag{84g}$$

$$=-\rho-\frac{(b-4a\rho)T_1}{2a}\left(\sin^2\left[\frac{1}{4}\cos^{-1}\left(1-\frac{32a^2(d-4a\rho^3+3b\rho^2-2c\rho)}{(b-4a\rho)^3T_1^2}\right)\right]\right) \tag{84h}$$

Similarly, the last two roots corresponding to j = 1 for both k = 0 and k = 1 are:

$$x_3=-\rho-\frac{(b-4a\rho)T_0}{4a}\left(1+\cos\left[\frac{1}{2}\cos^{-1}\left(1-\frac{32a^2(d-4a\rho^3+3b\rho^2-2c\rho)}{(b-4a\rho)^3T_0^2}\right)\right]\right) \tag{84i}$$

$$=-\rho-\frac{(b-4a\rho)T_0}{2a}\left(\cos^2\left[\frac{1}{4}\cos^{-1}\left(1-\frac{32\,a^2\left(d-4\,a\,\rho^3+3\,b\,\rho^2-2\,c\,\rho\right)}{(b-4\,a\,\rho)^3\,T_0^2}\right)\right]\right) \qquad (84\,k)$$

$$x_4=-\rho-\frac{(b-4a\rho)T_1}{4a}\left(1+\cos\left[\frac{1}{2}\cos^{-1}\left(1-\frac{32\,a^2\left(d-4\,a\,\rho^3+3\,b\,\rho^2-2\,c\,\rho\right)}{(b-4\,a\,\rho)^3\,T_1^2}\right)\right]\right) \qquad (84\,l)$$

$$=-\rho-\frac{(b-4a\rho)T_1}{2a}\left(\cos^2\left[\frac{1}{4}\cos^{-1}\left(1-\frac{32\,a^2\left(d-4\,a\,\rho^3+3\,b\,\rho^2-2\,c\,\rho\right)}{(b-4\,a\,\rho)^3\,T_1^2}\right)\right]\right) \qquad (84\,m)$$

Formula (84b) provides a solution to general quartic equation. However, this solution is only valid when the argument of the inverse cosine resides within the interval [-1, 1]. This situation is similar to that of cubic formula (40b). It implies analogous solution (84b) to cover roots within the intervals (-∞, -1] and [1, ∞), but in the form of the hyperbolic cosine and its inverse. Indeed, we have cosh(2cosh^{-1}(x)) = 2x²-1, which gives the same result as the composition of the double angle cosine with its inverse does. As a result, we obtain:

$$ax^4+bx^3+cx^2+dx+e=0 \quad (a\neq 0) \qquad (85a)$$

$$x_{j+1}=-\rho-\frac{(b-4a\rho)T_k}{4a}\left(1-\cosh\left[\frac{1}{2}\cosh^{-1}\left(1-\frac{32a^2\left(d+3b\rho^2-4a\rho^3-2c\rho\right)}{(b-4a\rho)^3T_k^2}\right)+i\pi j\right]\right),\,j=0,1. \qquad (85b)$$

where $\qquad (85)$

$$T_k=1-\cosh\left[\frac{1}{2}\cosh^{-1}\left(\frac{2\left(b^3+8a^2d-4abc\right)}{(b-4a\rho)^3}-1\right)+i\pi k\right],\,k=0,1. \qquad (85c)$$

$$\rho=\frac{16a^2e+b^2c+2abd-4ac^2}{3\left(b^3+8a^2d-4abc\right)}+\frac{2\sqrt{\left(4ac^2-2abd-b^2c-16a^2e\right)^2-3\left(b^3+8a^2d-4abc\right)\left(b^2d+8abe-4acd\right)}}{3\left(b^3+8a^2d-4abc\right)}*\cosh\left(\frac{1}{3}\cosh^{-1}(V)\right) \qquad (85d)$$

$$V=\frac{9\left(b^3+8a^2d-4abc\right)\left(4ac^2-2abd-b^2c-16a^2e\right)\left(b^2d+8abe-4acd\right)-2\left(4ac^2-2abd-b^2c-16a^2e\right)^3-27\left(b^3+8a^2d-4abc\right)^2\left(ad^2-b^2e\right)}{2\left(\left(4ac^2-2abd-b^2c-16a^2e\right)^2-3\left(b^3+8a^2d-4abc\right)\left(b^2d+8abe-4acd\right)\right)^{3/2}}$$

It is crucial to emphasize that we have the ability to express quartic formulas (84b), (84c), (85b) or (85c) in the radical form by using formulas (46a) or (46e) without loss of generality.

Analysis of solution form (84)

a. Transform T$_k$ into Radical Form

We proceed to transform the form of T$_k$ found in (84c) [or (85c)] into a radical form by using identity (1.3),

$$\cos\left(\frac{1}{2}\cos^{-1}(u)\right)=\frac{1}{\sqrt{2}}\sqrt{1+u},$$

as follows:

$$T_k=1-\cos\left[\frac{1}{2}\cos^{-1}\left(\frac{2\left(b^3+8\,a^2\,d-4\,a\,b\,c\right)}{\left(b-4\,a\,\rho\right)^3}-1\right)+\pi\,k\right],\quad k=0,1.$$

Putting k = 0 and k = 1, we have

$$T_{k=0,1}=1\mp\cos\left[\frac{1}{2}\cos^{-1}\left(1-\frac{8\,a\left(12\,a\,b\,\rho^2-16\,a^2\,\rho^3-3\,b^2\,\rho+b\,c-2\,a\,d\right)}{\left(b-4\,a\,\rho\right)^3}\right)\right]$$

$$=1\mp\cos\left[\frac{1}{2}\cos^{-1}\left(\frac{2\left(b^3+8\,a^2\,d-4\,a\,b\,c\right)}{\left(b-4\,a\,\rho\right)^3}-1\right)\right]$$

$$=1\mp\sqrt{\frac{b^3+8\,a^2\,d-4\,a\,b\,c}{\left(b-4\,a\,\rho\right)^3}}\qquad\qquad(85\,e)$$

b. Case of $b^3+8a^2d-4abc = 0$

We consider a case where solution formula (84b) or (85b) divides by zero. If $b^3+8a^2d-4abc$ equals zero, we observe that cubic equation (82) becomes the quadratic equation in ρ:

$$\left(4\,a\,c^2-b^2\,c-16\,a^2\,e-2\,a\,b\,d\right)\rho^2+\left(b^2\,d+8\,a\,b\,e-4\,a\,c\,d\right)\rho+a\,d^2-b^2\,e=0.\qquad(86)$$

We use traditional quadratic formula solve for ρ, which gives

$$\rho_{1,2}=\frac{-\left(b^2\,d+8\,a\,b\,e-4\,a\,c\,d\right)\pm\sqrt{\left(b^2\,d+8\,a\,b\,e-4\,a\,c\,d\right)^2-4\left(4\,a\,c^2-2\,a\,b\,d-b^2\,c-16\,a^2\,e\right)\left(a\,d^2-b^2\,e\right)}}{2\left(4\,a\,c^2-2\,a\,b\,d-b^2\,c-16\,a^2\,e\right)}\qquad(87)$$

$$=\frac{-\left(b^2\,d+8abe-4acd\right)\pm\sqrt{d^2b^4+8ba^2d^3+8adeb^3+64ed^2a^3+16aeb^2c^2-4ceb^4-4acb^2d^2-64bcdea^2}}{2\left(4ac^2-2abd-b^2c-16a^2e\right)},\qquad(88)$$

and solve for T_k from (85e) in which we get $T_{k=0,1} = 1$.

Next, by substituting $T_{k=0,1}$ into (84b), we obtain

$$x_1 = x_2 \text{ or } x_3 = x_4 = -\rho - \frac{b-4a\rho}{4a}\left(1 \mp \cos\left[\frac{1}{2}\cos^{-1}\left(1 - \frac{32a^2\left(d-4a\rho^3+3b\rho^2-2c\rho\right)}{\left(b-4a\rho\right)^3}\right)\right]\right).$$ (89)

(Note: This formula cannot be applied because b²-4aρ = 0.)

However, when we tested the formula (89) with a few examples, we found that the results were not calculated due to a division by zero error. This is because b²-4aρ is also equal to zero which contradicts the assumption stated in step 9. As a result, it appears that *formula (89) is invalid because (84b) or (85b) fails to handle in this case*. To prove this, we need to investigate why b²-4aρ is zero by taking the limit of (84b) as b³+8a²d-4abc approaches zero (or d approaches (4abc-b³)/(8a²)) as follows:

$$\rho = \lim_{d \to \frac{4abc-b^3}{8a^2}}\left\{\frac{16a^2e+b^2c+2abd-4ac^2}{3\left(b^3+8a^2d-4abc\right)} + \frac{\sqrt{\left(4ac^2-2abd-b^2c-16a^2e\right)^2 - 3\left(b^3+8a^2d-4abc\right)\left(b^2d+8abe-4acd\right)}}{3\left(b^3+8a^2d-4abc\right)}\right\}$$

By respectively taking the derivative of the numerator and denominator with respect to d using L'Hospital's Rule, we obtain

$$= \lim_{d \to \frac{4abc-b^3}{8a^2}}\left\{\frac{b}{12a} + \frac{2\left(4ac^2-2abd-b^2c-16a^2e\right)(-2ab)-3\left[\left(8a^2\right)\left(b^2d+8abe-4acd\right)+\left(b^3+8a^2d-4abc\right)\left(b^2-4ac\right)\right]}{48a^2\sqrt{\left(4ac^2-2abd-b^2c-16a^2e\right)^2 - 3\left(b^3+8a^2d-4abc\right)\left(b^2d+8abe-4acd\right)}}\right\}$$

Substituting the expression (b³+8a²d-4abc) by zero and simplifying further gives

$$= \lim_{d \to \frac{4abc-b^3}{8a^2}}\left\{\frac{b}{12a} - \frac{b\left(4ac^2-2abd-b^2c-16a^2e\right)+6a\left(b^2d+8abe-4acd\right)}{12a\left(4ac^2-2abd-b^2c-16a^2e\right)}\right\}$$

$$= \lim_{d \to \frac{4abc-b^3}{8a^2}}\left\{-\frac{\left(b^2d+8abe-4acd\right)}{2\left(4ac^2-2abd-b^2c-16a^2e\right)}\right\}$$

$$= \left\{-\frac{b^2\frac{4abc-b^3}{8a^2}+8abe-4ac\frac{4abc-b^3}{8a^2}}{2\left(4ac^2-2ab\frac{4abc-b^3}{8a^2}-b^2c-16a^2e\right)}\right\}$$

$$= \frac{b\left(b^4-8ab^2c-64a^3e+16a^2c^2\right)}{4a\left(b^4-8ab^2c-64a^3e+16a^2c^2\right)} = \frac{b}{4a},$$ (90)

which leads to b-4aρ equals zero. Then the proof is complete. How do we find the solution to equation (70) in this case?

c. Case b²-4aρ = 0 when b³+8a²d-4abc = 0

We now face the challenge of finding solution to the quartic equation when both $b^3+8a^2d-4abc$ and $b^2-4a\rho$ approach zero. How do we find solution to the quartic equation in this case? To address this, we need to go back to (77a) and (77b) at step 8 which defines parameter ρ. If substituting $\rho = -b/(4a)$ into (77a) is the equivalent to substituting $x = u - b/(4a)$ into (84a), then the result is shown below:

$$a u^4 + \left(c - \frac{3 b^2}{8 a}\right) u^2 + \left(\frac{b^3 + 8 a^2 d - 4 a b c}{8 a^2}\right) u + e - \frac{3 b^4}{256 a^3} + \frac{b^2 c}{16 a^2} - \frac{b d}{4 a} \tag{91}$$

Under the condition $(b^3+8a^2d-4abc) = 0$, equation (91) reduces to a biquadratic equation in u,

$$a u^4 + \left(c - \frac{3 b^2}{8 a}\right) u^2 + e - \frac{3 b^4}{256 a^3} + \frac{b^2 c}{16 a^2} - \frac{b d}{4 a}. \tag{92}$$

Using traditional quadratic formula to solve for u gives

$$u^2_{1,2} = -\frac{1}{16 a^2}\left(8 a c - 3 b^2\right) \pm \frac{1}{16 a^2} \sqrt{\left(8 a c - 3 b^2\right)^2 - \left(5 b^4 + 256 a^3 e - 16 a b^2 c\right)}$$

$$= \frac{1}{16 a^2}\left(3 b^2 - 8 a c \pm 2 \sqrt{b^4 + 16 a^2 c^2 - 64 a^3 e - 8 a b^2 c}\right)$$

Substituting each of u with $x + b/(4a)$, then taking square root of both sides and re-arranging gives four roots to the quartic equation, respectively.

$$\begin{cases} x_{1,2} = -\frac{b}{4 a} \pm \frac{1}{4 a} \sqrt{3 b^2 - 8 a c + 2 \sqrt{b^4 + 16 a^2 c^2 - 64 a^3 e - 8 a b^2 c}} \\ x_{3,4} = -\frac{b}{4 a} \pm \frac{1}{4 a} \sqrt{3 b^2 - 8 a c - 2 \sqrt{b^4 + 16 a^2 c^2 - 64 a^3 e - 8 a b^2 c}}. \end{cases} \tag{93}$$

If the expression $3 b^2 - 8 a c + 2 \sqrt{b^4 + 16 a^2 c^2 - 64 a^3 e - 8 a b^2 c}$ equals zero, then formula (93) is simplified to $-b/(4a)$, which gives a repeated root to the equation.

Converting (84) to a Depressed Quartic

A depressed quartic is a quartic equation that lacks a cubic term. It can be expressed in the following form:

$$a x^4 + c x^2 + d x + e = 0 \quad (a \neq 0). \tag{94}$$

Through the substitution x with x-b/(4a) in (84) we can transform the quartic equation into a depressed form. An alternative approach is used by putting b = 0 in (84) that leads to the same result, according to our analysis. We present below the alternative approach. Readers may refer to the Appendix [2-A-12] for the substitution approach.

To derive (94), we plug $T_{k=0,1}$ found from (84e) in (84), which gives

$$x_{j+1} = -\rho - \frac{(b-4a\rho)^{3/2} \mp \sqrt{b^3 + 8a^2 d - 4abc}}{4a\sqrt{b-4a\rho}} \left(1 - \cos\left[\frac{1}{2}\cos^{-1}\left(1 - \frac{32a^2(d-4a\rho^3 + 3b\rho^2 - 2c\rho)}{\left((b-4a\rho)^{3/2} \mp \sqrt{b^3 + 8a^2 d - 4abc}\right)^2} \right) + \pi j \right] \right), j = 0,1. \tag{95}$$

Formula (95) indicates there are two roots for each value of j. Hence, it explicitly gives four roots in total as follows:

$$\left\{ \begin{aligned} x_{1,2} &= -\rho - \frac{(b-4a\rho)^{3/2} \mp \sqrt{b^3 + 8a^2 d - 4abc}}{4a\sqrt{b-4a\rho}} \left(1 - \cos\left[\frac{1}{2}\cos^{-1}\left(1 - \frac{32a^2(d-4a\rho^3 + 3b\rho^2 - 2c\rho)}{\left((b-4a\rho)^{3/2} \mp \sqrt{b^3 + 8a^2 d - 4abc}\right)^2} \right) \right] \right) & (96a) \\[4mm] x_{3,4} &= -\rho - \frac{(b-4a\rho)^{3/2} \mp \sqrt{b^3 + 8a^2 d - 4abc}}{4a\sqrt{b-4a\rho}} \left(1 + \cos\left[\frac{1}{2}\cos^{-1}\left(1 - \frac{32a^2(d-4a\rho^3 + 3b\rho^2 - 2c\rho)}{\left((b-4a\rho)^{3/2} \mp \sqrt{b^3 + 8a^2 d - 4abc}\right)^2} \right) \right] \right) & (96b) \end{aligned} \right. \tag{96}$$

Now, plugging b = 0 in (96) results in:

$$\left\{ \begin{aligned} x_{1,2} &= -\rho - \frac{4\sqrt{a}(-\rho)^{3/2} \mp \sqrt{2d}}{4\sqrt{-a\rho}} \left(1 - \cos\left[\frac{1}{2}\cos^{-1}\left(1 - \frac{8(d-4a\rho^3 - 2c\rho)}{\left(4\sqrt{a}(-\rho)^{3/2} \mp \sqrt{2d}\right)^2} \right) \right] \right) & (97a) \\[4mm] x_{3,4} &= -\rho - \frac{4\sqrt{a}(-\rho)^{3/2} \mp \sqrt{2d}}{4\sqrt{-a\rho}} \left(1 + \cos\left[\frac{1}{2}\cos^{-1}\left(1 - \frac{8(d-4a\rho^3 - 2c\rho)}{\left(4\sqrt{a}(-\rho)^{3/2} \mp \sqrt{2d}\right)^2} \right) \right] \right) & (97b) \\[4mm] \text{where} \\[2mm] \rho &= \frac{4ae-c^2}{6ad} + \frac{\sqrt{(c^2-4ae)^2 + 6acd^2}}{3ad} * \cos\left(\frac{1}{3}\cos^{-1}\left(\frac{18acd^2(4ae-c^2) + 2(4ae-c^2)^3 - 27a^2 d^4}{2\left((c^2-4ae)^2 + 6acd^2\right)^{3/2}} \right) \right). & (97c) \end{aligned} \right. \tag{97}$$

Replacing ρ with -ρ in (97) to remove the minus sign out of the square root gives solutions to depressed quartic (94):

$$x_{1,2}=-\rho-\frac{4\sqrt{a}(\rho)^{3/2}\mp\sqrt{2d}}{4\sqrt{a\rho}}\left(1-\cos\left[\frac{1}{2}\cos^{-1}\left(1-\frac{8(d+4a\rho^3+2c\rho)}{\left(4\sqrt{a}(\rho)^{3/2}\mp\sqrt{2d}\right)^2}\right)\right]\right),\quad (98a)$$

$$x_{3,4}=-\rho-\frac{4\sqrt{a}(\rho)^{3/2}\mp\sqrt{2d}}{4\sqrt{a\rho}}\left(1+\cos\left[\frac{1}{2}\cos^{-1}\left(1-\frac{8(d+4a\rho^3+2c\rho)}{\left(4\sqrt{a}(\rho)^{3/2}\mp\sqrt{2d}\right)^2}\right)\right]\right),\quad (98b) \quad (98)$$

where

$$\rho=-\frac{4ae-c^2}{6ad}-\frac{\sqrt{(c^2-4ae)^2+6acd^2}}{3ad}*\cos\left(\frac{1}{3}\cos^{-1}\left(\frac{18acd^2(4ae-c^2)+2(4ae-c^2)^3-27a^2d^4}{2\left((c^2-4ae)^2+6acd^2\right)^{3/2}}\right)\right). \quad (98c)$$

Moreover, we do the same process for (98) to derive the similar formula, but in hyperbolic form. Please refer to the Appendix [2-A-13] for the depressed quartic with the coefficients renamed in the form: $ax^4+bx^2+cx+d=0$.

Transform Quartic Formula into Radical Form

We express quartic formula (84b) or (85b) to a radical form. By applying cosine formula (1.3) to (96a), we have:

$$x_{1,2}=-\rho-\frac{(b-4a\rho)^{3/2}\mp\sqrt{b^3+8a^2d-4abc}}{4a\sqrt{b-4a\rho}}\left(1-\frac{1}{\sqrt{2}}\sqrt{2-\frac{32a^2(d-4a\rho^3+3b\rho^2-2c\rho)}{\left((b-4a\rho)^{3/2}\mp\sqrt{b^3+8a^2d-4abc}\right)^2}}\right)$$

$$=-\rho-\frac{(b-4a\rho)^{3/2}\mp\sqrt{b^3+8a^2d-4abc}}{4a\sqrt{b-4a\rho}}\left(1-\left(\frac{\sqrt{\left((b-4a\rho)^{3/2}\mp\sqrt{b^3+8a^2d-4abc}\right)^2-16a^2(d-4a\rho^3+3b\rho^2-2c\rho)}}{(b-4a\rho)^{3/2}\mp\sqrt{b^3+8a^2d-4abc}}\right)\right)$$

$$=-\rho-\frac{(b-4a\rho)^{3/2}\mp\sqrt{b^3+8a^2d-4abc}-\sqrt{\left((b-4a\rho)^{3/2}\mp\sqrt{b^3+8a^2d-4abc}\right)^2-16a^2(d-4a\rho^3+3b\rho^2-2c\rho)}}{4a\sqrt{b-4a\rho}}$$

Similarly, we can get the formulas for x_3 and x_4,

$$x_{3,4}=-\rho-\frac{(b-4a\rho)^{3/2}\mp\sqrt{b^3+8a^2d-4abc}+\sqrt{\left((b-4a\rho)^{3/2}\mp\sqrt{b^3+8a^2d-4abc}\right)^2-16a^2(d-4a\rho^3+3b\rho^2-2c\rho)}}{4a\sqrt{b-4a\rho}}$$

Thus, the general quartic formula expressed in radical form is

$$\left\{ \begin{array}{l} x_{1,2}=-\rho-\dfrac{(b-4a\rho)^{3/2}\mp\sqrt{b^3+8a^2d-4abc}-\sqrt{\left((b-4a\rho)^{3/2}\mp\sqrt{b^3+8a^2d-4abc}\right)^2-16a^2(d-4a\rho^3+3b\rho^2-2c\rho)}}{4a\sqrt{b-4a\rho}} \\[4mm] x_{3,4}=-\rho-\dfrac{(b-4a\rho)^{3/2}\mp\sqrt{b^3+8a^2d-4abc}+\sqrt{\left((b-4a\rho)^{3/2}\mp\sqrt{b^3+8a^2d-4abc}\right)^2-16a^2(d-4a\rho^3+3b\rho^2-2c\rho)}}{4a\sqrt{b-4a\rho}}. \end{array} \right. \tag{99}$$

where p is the real root of cubic equation getting from cubic formula (84b) or (85b). Formula (85b) can be converted in radical form using (1.3.3a) at n = 3.

Next, we examine two summaries. Firstly, we present an overview of solution to the quartic formula expressed in terms of radicals, where the cubic formula is also expressed in radical form without involving $T_{k=0,1}$. Secondly, we provide a summary of quartic formulas in terms of trigonometric and hyperbolic trigonometric functions, involving both the quadratic and cubic formulas. Then we demonstrate an example of using the quartic formula to solve a quartic equation.

A. Summary of Solution to Quartic Equation in Radical Form without T_k

Quartic equation: $a x^4+b x^3+c x^2+d x+e=0 \ (a\neq 0)$

Define

$$O=b^3+8 a^2 d-4 a b c \tag{99 a}$$

$$M=b-4 a \rho \tag{99 b}$$

$$N=16 a^2\left(3 b \rho^2+d-4 a \rho^3-2 c \rho\right) \tag{99 c}$$

The solution in radicals to the quartic equation is given by formula (99):

$$x_1=-\rho-\dfrac{M^{3/2}-\sqrt{O}-\sqrt{\left(M^{3/2}-\sqrt{O}\right)^2-N}}{4 a \sqrt{M}}, \tag{99 d}$$

$$x_2=-\rho-\dfrac{M^{3/2}+\sqrt{O}-\sqrt{\left(M^{3/2}+\sqrt{O}\right)^2-N}}{4 a \sqrt{M}}, \tag{99 e}$$

$$x_3=-\rho-\dfrac{M^{3/2}-\sqrt{O}+\sqrt{\left(M^{3/2}-\sqrt{O}\right)^2-N}}{4 a \sqrt{M}}, \tag{99 f}$$

$$x_4 = -\rho - \frac{M^{3/2} + \sqrt{O} + \sqrt{\left(M^{3/2} + \sqrt{O}\right)^2 - N}}{4\,a\sqrt{M}}. \qquad (99\,g)$$

where ρ is given by (83)

$$\rho = \begin{cases} -\dfrac{P}{Q} + \dfrac{2\sqrt{D}}{Q}\cos\left(\dfrac{1}{3}\cos^{-1}(V)\right), & D \geq 0,\ -1 \leq V \leq 1, \\[3mm] -\dfrac{P}{Q} + \dfrac{2\sqrt{D}}{Q}\cosh\left(\dfrac{1}{3}\cosh^{-1}(V)\right), & D \geq 0,\ V \geq 1, \\[3mm] -\dfrac{P}{Q} - \dfrac{2\sqrt{D}}{Q}\cosh\left(\dfrac{1}{3}\cosh^{-1}(|V|)\right), & D \geq 0,\ V \leq -1, \\[3mm] -\dfrac{P}{Q} + \dfrac{2\sqrt{|D|}}{Q}\sinh\left(\dfrac{1}{3}\sinh^{-1}\left(\dfrac{3QPR - 2P^3 - 3Q^2 S}{2(|D|)^{3/2}}\right)\right), & D < 0,\ V \in \mathbb{R} \end{cases} \qquad (99\,h)$$

or ρ can be given in radical form:

$$\rho = -\frac{P}{Q} + \frac{\sqrt{D}}{Q}\left[\left(V - i\sqrt{1 - V^2}\right)^{1/3} + \left(V - i\sqrt{1 - V^2}\right)^{-1/3}\right], \qquad D \geq 0, |V| \leq 1 \ (99\,i)$$

$$\rho = -\frac{P}{Q} + \frac{\sqrt{D}}{Q}\left[\left(V + \sqrt{V^2 - 1}\right)^{1/3} + \left(V + \sqrt{V^2 - 1}\right)^{-1/3}\right], \qquad D \geq 0, V \geq 1 \ (99\,k)$$

$$\rho = -\frac{P}{Q} - \frac{\sqrt{D}}{Q}\left[\left(|V| + \sqrt{V^2 - 1}\right)^{1/3} + \left(|V| + \sqrt{V^2 - 1}\right)^{-1/3}\right], \qquad D \geq 0, V \leq -1 \ (99\,k)$$

$$\rho = -\frac{P}{Q} + \frac{\sqrt{|D|}}{iQ}\left[\left(V_1 + \sqrt{V_1^2 + 1}\right)^{1/3} - \left(V_1 + \sqrt{V_1^2 + 1}\right)^{-1/3}\right], \qquad D < 0, V_1 \in \mathbb{R} \ (99\,l)$$

(Using formulas (46e), (46a) and (46b) transform all in (99h).)

where

$$P = 4\,a\,c^2 - 16\,a^2 e - b^2 c - 2\,a\,b\,d, \qquad (99\,m)$$

$$Q = 3\left(b^3 + 8\,a^2 d - 4\,a\,b\,c\right), \qquad (99\,n)$$

$$R = b^2 d + 8\,a\,b\,e - 4\,a\,c\,d, \qquad (99\,o)$$

$$S = a\,d^2 - b^2 e \qquad (99\,p)$$

$$D = P^2 - QR \qquad (99\,p)$$

and

$$V = \frac{3QPR - 2P^3 - 3Q^2 S}{2(D)^{3/2}}. \qquad (99\,q)$$

Note that only V_1 in (99l) must be defined as

$$V_1 = \frac{3QPR - 2P^3 - 3Q^2 S}{2(|D|)^{3/2}}. \qquad (99\,s)$$

Special Case

If M = 0, the solution to the quartic equation are given by (93).

The following summary presents analysis of quartic formula (84b) or (85b) and gives closed-form solution, specifically designed suitable for simpler calculator. However, the solution maybe lengthy and our aim is to represent the solution formulas expressed in terms of cosine and hyperbolic cosine functions in order to investigate their relationship with other trigonometric and hyperbolic trigonometric functions, with the potential benefit of inspiring the creation of new elementary functions later in this chapter.

B. Summary *of Solution to Quartic Equation with* T_k

Quartic equation: $a x^4 + b x^3 + c x^2 + d x + e = 0 \quad (a \neq 0)$

Define

$$M = \frac{2(b^3 + 8a^2 d - 4abc)}{(b - 4a\rho)^3} - 1, \quad (b \neq 4a\rho),$$

$$N_k = 1 - \frac{32 a^2 (d - 4a\rho^3 + 3b\rho^2 - 2c\rho)}{(b - 4a\rho)^3 T_k^2},$$

where ρ is given by cubic formula (83), and T_k is given by either (84c) or (85c).

Analysis of M and T_k:

- If $-1 \leq M \leq 1$,

$$T_{k=0,1}=1-\cos\left(\frac{1}{2}\cos^{-1}(M)+\pi k\right)=\begin{cases}T_0=1-\cos\left(\frac{1}{2}\cos^{-1}(M)\right)=2\sin^2\left(\frac{1}{4}\cos^{-1}(M)\right)\\T_1=1+\cos\left(\frac{1}{2}\cos^{-1}(M)\right)=2\cos^2\left(\frac{1}{4}\cos^{-1}(M)\right).\end{cases}$$

- If $M \geq 1$,

$$T_{k=0,1}=1-\cosh\left(\frac{1}{2}\cosh^{-1}(M)+i\pi k\right)=\begin{cases}T_0=1-\cosh\left(\frac{1}{2}\cosh^{-1}(M)\right)=-2\sinh^2\left(\frac{1}{4}\cosh^{-1}(M)\right)\\T_1=1+\cosh\left(\frac{1}{2}\cosh^{-1}(M)\right)=2\cosh^2\left(\frac{1}{4}\cosh^{-1}(M)\right).\end{cases}$$

- If $M < -1$,

$$T_{k=0,1}=1+i\sinh\left(\frac{1}{2}\cosh^{-1}(|M|)-i\pi k\right)=\begin{cases}T_0=1+i\sinh\left(\frac{1}{2}\cosh^{-1}(|M|)\right),\\T_1=1-i\sinh\left(\frac{1}{2}\cosh^{-1}(|M|)\right),\end{cases}$$

where i is the imaginary unit.

Analysis of M and N_k in Solution Formula (84b) and (85b)

1. Case $-1 < M \leq 1$:

a. $-1 \leq N_k \leq 1$:

The equation has four distinct real roots, given by formula (84b):

$$\begin{cases}x_{k+1}=-\rho-\dfrac{(b-4a\rho)T_k}{4a}\left(1-\cos\left[\frac{1}{2}\cos^{-1}(N_k)\right]\right),k=0,1\\x_{k+3}=-\rho-\dfrac{(b-4a\rho)T_k}{4a}\left(1+\cos\left[\frac{1}{2}\cos^{-1}(N_k)\right]\right),k=0,1,\end{cases}$$

The roots are explicitly expressed in closed form:

$$\begin{cases} x_1=-\rho-\dfrac{(b-4a\rho)T_0}{4a}\left(1-\cos\left[\dfrac{1}{2}\cos^{-1}(N_0)\right]\right)=-\rho-\left(\dfrac{b}{a}-4\rho\right)\sin^2\left(\dfrac{1}{4}\cos^{-1}(M)\right)\sin^2\left(\dfrac{1}{4}\cos^{-1}(N_0)\right) \\[4mm] x_2=-\rho-\dfrac{(b-4a\rho)T_1}{4a}\left(1-\cos\left[\dfrac{1}{2}\cos^{-1}(N_1)\right]\right)=-\rho-\left(\dfrac{b}{a}-4\rho\right)\cos^2\left(\dfrac{1}{4}\cos^{-1}(M)\right)\sin^2\left(\dfrac{1}{4}\cos^{-1}(N_1)\right) \\[4mm] x_3=-\rho-\dfrac{(b-4a\rho)T_0}{4a}\left(1+\cos\left[\dfrac{1}{2}\cos^{-1}(N_0)\right]\right)=-\rho-\left(\dfrac{b}{a}-4\rho\right)\sin^2\left(\dfrac{1}{4}\cos^{-1}(M)\right)\cos^2\left(\dfrac{1}{4}\cos^{-1}(N_0)\right) \\[4mm] x_4=-\rho-\dfrac{(b-4a\rho)T_1}{4a}\left(1+\cos\left[\dfrac{1}{2}\cos^{-1}(N_1)\right]\right)=-\rho-\left(\dfrac{b}{a}-4\rho\right)\cos^2\left(\dfrac{1}{4}\cos^{-1}(M)\right)\cos^2\left(\dfrac{1}{4}\cos^{-1}(N_1)\right). \end{cases}$$

b. $N_k \geq 1$:

The equation has four real roots, given by formula (85b):

$$\begin{cases} x_{k+1}=-\rho-\dfrac{(b-4a\rho)T_k}{4a}\left(1-\cosh\left[\dfrac{1}{2}\cosh^{-1}(N_k)\right]\right),\ k=0,1 \\[4mm] x_{k+3}=-\rho-\dfrac{(b-4a\rho)T_k}{4a}\left(1+\cosh\left[\dfrac{1}{2}\cosh^{-1}(N_k)\right]\right),\ k=0,1, \end{cases}$$

The roots are explicitly expressed in closed form:

$$\begin{cases} x_1=-\rho-\dfrac{(b-4a\rho)T_0}{4a}\left(1-\cosh\left[\dfrac{1}{2}\cosh^{-1}(N_0)\right]\right)=-\rho+\left(\dfrac{b}{a}-4\rho\right)\sin^2\left(\dfrac{1}{4}\cos^{-1}(M)\right)\sinh^2\left(\dfrac{1}{4}\cosh^{-1}(N_0)\right) \\[4mm] x_2=-\rho-\dfrac{(b-4a\rho)T_1}{4a}\left(1-\cosh\left[\dfrac{1}{2}\cosh^{-1}(N_1)\right]\right)=-\rho+\left(\dfrac{b}{a}-4\rho\right)\cos^2\left(\dfrac{1}{4}\cosh^{-1}(M)\right)\sinh^2\left(\dfrac{1}{4}\cosh^{-1}(N_1)\right) \\[4mm] x_3=-\rho-\dfrac{(b-4a\rho)T_0}{4a}\left(1+\cosh\left[\dfrac{1}{2}\cosh^{-1}(N_0)\right]\right)=-\rho-\left(\dfrac{b}{a}-4\rho\right)\sin^2\left(\dfrac{1}{4}\cos^{-1}(M)\right)\cosh^2\left(\dfrac{1}{4}\cosh^{-1}(N_0)\right) \\[4mm] x_4=-\rho-\dfrac{(b-4a\rho)T_1}{4a}\left(1+\cosh\left[\dfrac{1}{2}\cosh^{-1}(N_1)\right]\right)=-\rho-\left(\dfrac{b}{a}-4\rho\right)\cos^2\left(\dfrac{1}{4}\cos^{-1}(M)\right)\cosh^2\left(\dfrac{1}{4}\cosh^{-1}(N_1)\right). \end{cases}$$

c. $N_k < -1$:

The equation has two pairs of the complex conjugate roots, given by formula (85b):

$$\begin{cases} x_{k+1}=-\rho-\dfrac{(b-4a\rho)T_k}{4a}\left(1+i\sinh\left[\dfrac{1}{2}\cosh^{-1}(|N_k|)\right]\right),\ k=0,1 \\[4mm] x_{k+3}=-\rho-\dfrac{(b-4a\rho)T_k}{4a}\left(1-i\sinh\left[\dfrac{1}{2}\cosh^{-1}(|N_k|)\right]\right),\ k=0,1, \end{cases}$$

The roots are explicitly expressed in closed form:

$$\left\{\begin{aligned}
x_1 &= -\rho - \frac{(b-4a\rho)T_0}{4a}\left(1 + i\sinh\left[\frac{1}{2}\cosh^{-1}(|N_0|)\right]\right) = -\rho - \left(\frac{b}{2a} - 2\rho\right)\sin^2\left(\frac{1}{4}\cos^{-1}(M)\right)\left(1 + i\sinh\left[\frac{1}{2}\cosh^{-1}(|N_0|)\right]\right) \\
x_2 &= -\rho - \frac{(b-4a\rho)T_1}{4a}\left(1 + i\sinh\left[\frac{1}{2}\cosh^{-1}(|N_1|)\right]\right) = -\rho - \left(\frac{b}{2a} - 2\rho\right)\cos^2\left(\frac{1}{4}\cos^{-1}(M)\right)\left(1 + i\sinh\left[\frac{1}{2}\cosh^{-1}(|N_1|)\right]\right) \\
x_3 &= -\rho - \frac{(b-4a\rho)T_0}{4a}\left(1 - i\sinh\left[\frac{1}{2}\cosh^{-1}(|N_0|)\right]\right) = -\rho - \left(\frac{b}{2a} - 2\rho\right)\sin^2\left(\frac{1}{4}\cos^{-1}(M)\right)\left(1 - i\sinh\left[\frac{1}{2}\cosh^{-1}(|N_0|)\right]\right) \\
x_4 &= -\rho - \frac{(b-4a\rho)T_1}{4a}\left(1 - i\sinh\left[\frac{1}{2}\cosh^{-1}(|N_1|)\right]\right) = -\rho - \left(\frac{b}{2a} - 2\rho\right)\cos^2\left(\frac{1}{4}\cos^{-1}(M)\right)\left(1 - i\sinh\left[\frac{1}{2}\cosh^{-1}(|N_1|)\right]\right).
\end{aligned}\right.$$

2. Case M ≥ 1:

a. -1 ≤ N_k ≤ 1:

The equation has four distinct real roots, provided by formula (84b):

$$\left\{\begin{aligned}
x_{k+1} &= -\rho - \frac{(b-4a\rho)T_k}{4a}\left(1 - \cos\left[\frac{1}{2}\cos^{-1}(N_k)\right]\right), \quad k=0,1 \\
x_{k+3} &= -\rho - \frac{(b-4a\rho)T_k}{4a}\left(1 + \cos\left[\frac{1}{2}\cos^{-1}(N_k)\right]\right), \quad k=0,1.
\end{aligned}\right.$$

The roots are explicitly expressed in closed form:

$$\left\{\begin{aligned}
x_1 &= -\rho - \frac{(b-4a\rho)T_0}{4a}\left(1 - \cos\left[\frac{1}{2}\cos^{-1}(N_0)\right]\right) = -\rho + \left(\frac{b}{a} - 4\rho\right)\sinh^2\left(\frac{1}{4}\cosh^{-1}(M)\right)\sin^2\left(\frac{1}{4}\cos^{-1}(N_0)\right) \\
x_2 &= -\rho - \frac{(b-4a\rho)T_1}{4a}\left(1 - \cos\left[\frac{1}{2}\cos^{-1}(N_1)\right]\right) = -\rho - \left(\frac{b}{a} - 4\rho\right)\cosh^2\left(\frac{1}{4}\cosh^{-1}(M)\right)\sin^2\left(\frac{1}{4}\cos^{-1}(N_1)\right) \\
x_3 &= -\rho - \frac{(b-4a\rho)T_0}{4a}\left(1 + \cos\left[\frac{1}{2}\cos^{-1}(N_0)\right]\right) = -\rho + \left(\frac{b}{a} - 4\rho\right)\sinh^2\left(\frac{1}{4}\cosh^{-1}(M)\right)\cos^2\left(\frac{1}{4}\cos^{-1}(N_0)\right) \\
x_4 &= -\rho - \frac{(b-4a\rho)T_1}{4a}\left(1 + \cos\left[\frac{1}{2}\cos^{-1}(N_1)\right]\right) = -\rho - \left(\frac{b}{a} - 4\rho\right)\cosh^2\left(\frac{1}{4}\cosh^{-1}(M)\right)\cos^2\left(\frac{1}{4}\cos^{-1}(N_1)\right)
\end{aligned}\right.$$

b. N_k ≥ 1:

The equation has four distinct real roots via formula (85b):

$$\left\{\begin{aligned}
x_{k+1} &= -\rho - \frac{(b-4a\rho)T_k}{4a}\left(1 - \cosh\left[\frac{1}{2}\cosh^{-1}(N_k)\right]\right), \quad k=0,1 \\
x_{k+3} &= -\rho - \frac{(b-4a\rho)T_k}{4a}\left(1 + \cosh\left[\frac{1}{2}\cosh^{-1}(N_k)\right]\right), \quad k=0,1.
\end{aligned}\right.$$

The roots are explicitly expressed in closed form:

$$\begin{cases}
x_1 = -\rho - \dfrac{(b-4a\rho)T_0}{4a}\left(1-\cosh\left[\dfrac{1}{2}\cosh^{-1}(N_0)\right]\right) = -\rho - \left(\dfrac{b}{a}-4\rho\right)\sinh^2\left(\dfrac{1}{4}\cosh^{-1}(M)\right)\sinh^2\left(\dfrac{1}{4}\cosh^{-1}(N_0)\right) \\[2mm]
x_2 = -\rho - \dfrac{(b-4a\rho)T_1}{4a}\left(1-\cosh\left[\dfrac{1}{2}\cosh^{-1}(N_1)\right]\right) = -\rho + \left(\dfrac{b}{a}-4\rho\right)\cosh^2\left(\dfrac{1}{4}\cosh^{-1}(M)\right)\sinh^2\left(\dfrac{1}{4}\cosh^{-1}(N_1)\right) \\[2mm]
x_3 = -\rho - \dfrac{(b-4a\rho)T_0}{4a}\left(1+\cosh\left[\dfrac{1}{2}\cosh^{-1}(N_0)\right]\right) = -\rho + \left(\dfrac{b}{a}-4\rho\right)\sinh^2\left(\dfrac{1}{4}\cosh^{-1}(M)\right)\cosh^2\left(\dfrac{1}{4}\cosh^{-1}(N_0)\right) \\[2mm]
x_4 = -\rho - \dfrac{(b-4a\rho)T_1}{4a}\left(1+\cosh\left[\dfrac{1}{2}\cosh^{-1}(N_1)\right]\right) = -\rho - \left(\dfrac{b}{a}-4\rho\right)\cosh^2\left(\dfrac{1}{4}\cosh^{-1}(M)\right)\cosh^2\left(\dfrac{1}{4}\cosh^{-1}(N_1)\right).
\end{cases}$$

c. $N_k < -1$:

The equation has two pairs of the complex conjugate roots, given by formula (85b):

$$\begin{cases}
x_{k+1} = -\rho - \dfrac{(b-4a\rho)T_k}{4a}\left(1+i\sinh\left[\dfrac{1}{2}\cosh^{-1}(|N_k|)\right]\right), \quad k=0,1 \\[2mm]
x_{k+3} = -\rho - \dfrac{(b-4a\rho)T_k}{4a}\left(1-i\sinh\left[\dfrac{1}{2}\cosh^{-1}(|N_k|)\right]\right), \quad k=0,1,
\end{cases}$$

The roots are explicitly expressed in closed form:

$$\begin{cases}
x_1 = -\rho - \dfrac{(b-4a\rho)T_0}{4a}\left(1+i\sinh\left[\dfrac{1}{2}\cosh^{-1}(|N_0|)\right]\right) = -\rho + \left(\dfrac{b}{2a}-2\rho\right)\sinh^2\left(\dfrac{1}{4}\cosh^{-1}(M)\right)\left(1+i\sinh\left[\dfrac{1}{2}\cosh^{-1}(|N_0|)\right]\right) \\[2mm]
x_2 = -\rho - \dfrac{(b-4a\rho)T_1}{4a}\left(1+i\sinh\left[\dfrac{1}{2}\cosh^{-1}(|N_1|)\right]\right) = -\rho - \left(\dfrac{b}{2a}-2\rho\right)\cosh^2\left(\dfrac{1}{4}\cosh^{-1}(M)\right)\left(1+i\sinh\left[\dfrac{1}{2}\cosh^{-1}(|N_1|)\right]\right) \\[2mm]
x_3 = -\rho - \dfrac{(b-4a\rho)T_0}{4a}\left(1-i\sinh\left[\dfrac{1}{2}\cosh^{-1}(|N_0|)\right]\right) = -\rho + \left(\dfrac{b}{2a}-2\rho\right)\sinh^2\left(\dfrac{1}{4}\cosh^{-1}(M)\right)\left(1-i\sinh\left[\dfrac{1}{2}\cosh^{-1}(|N_0|)\right]\right) \\[2mm]
x_4 = -\rho - \dfrac{(b-4a\rho)T_1}{4a}\left(1-i\sinh\left[\dfrac{1}{2}\cosh^{-1}(|N_1|)\right]\right) = -\rho - \left(\dfrac{b}{2a}-2\rho\right)\cosh^2\left(\dfrac{1}{4}\cosh^{-1}(M)\right)\left(1-i\sinh\left[\dfrac{1}{2}\cosh^{-1}(|N_1|)\right]\right).
\end{cases}$$

3. Case $M < -1$:

The solution to the quartic equation is given by (84b)

if $-1 \leq N_k \leq 1$,

$$\begin{cases} x_{k+1}=-\rho-\dfrac{(b-4a\rho)T_k}{4a}\left(1-\cos\left[\dfrac{1}{2}\cos^{-1}(N_k)\right]\right),\,k=0,1 \\[3mm] x_{k+3}=-\rho-\dfrac{(b-4a\rho)T_k}{4a}\left(1+\cos\left[\dfrac{1}{2}\cos^{-1}(N_k)\right]\right),\,k=0,1 \end{cases}$$

or by (85b) if $N_k \geq 1$,

$$\begin{cases} x_{k+1}=-\rho-\dfrac{(b-4a\rho)T_k}{4a}\left(1-\cosh\left[\dfrac{1}{2}\cosh^{-1}(N_k)\right]\right),\,k=0,1 \\[3mm] x_{k+3}=-\rho-\dfrac{(b-4a\rho)T_k}{4a}\left(1+\cosh\left[\dfrac{1}{2}\cosh^{-1}(N_k)\right]\right),\,k=0,1. \end{cases}$$

if $N_k \leq -1$,

$$\begin{cases} x_{k+1}=-\rho-\dfrac{(b-4a\rho)T_k}{4a}\left(1+i\sinh\left[\dfrac{1}{2}\cosh^{-1}(|N_k|)\right]\right),\,k=0,1 \\[3mm] x_{k+3}=-\rho-\dfrac{(b-4a\rho)T_k}{4a}\left(1-i\sinh\left[\dfrac{1}{2}\cosh^{-1}(|N_k|)\right]\right),\,k=0,1. \end{cases}$$

4. Case M = -1 or (b³ + 8a²d − 4abc) = 0:

As $(b^3 + 8a^2d - 4abc)$ approaches zero, then ρ approaches $b/(4a)$ or $(b - 4a\rho)$ tends to zero. This leads to the quartic equation becoming the biquadratic equation. The solution to the equation is given by (93):

$$x_{1,2}=-\frac{b}{4a}\pm\frac{1}{4a}\sqrt{3b^2-8ac+2\sqrt{b^4+16a^2c^2-64a^3e-8ab^2c}}$$

and

$$x_{3,4}=-\frac{b}{4a}\pm\frac{1}{4a}\sqrt{3b^2-8ac-2\sqrt{b^4+16a^2c^2-64a^3e-8ab^2c}}.$$

Notice that if M is defined as

$$M=\frac{(b^3+8a^2d-4abc)}{(b-4a\rho)^3},$$

then the domain of M can be determined in the following cases:

- $$-1\leq\frac{2(b^3+8a^2d-4abc)}{(b-4a\rho)^3}-1\leq1 \Leftrightarrow 0\leq\frac{b^3+8a^2d-4abc}{(b-4a\rho)^3}\leq1 \Leftrightarrow 0\leq M\leq2$$

- $$\frac{2\left(b^3+8a^2d-4abc\right)}{\left(b-4a\rho\right)^3}-1\geq1\Leftrightarrow\frac{b^3+8a^2d-4abc}{\left(b-4a\rho\right)^3}\geq2\Leftrightarrow M\geq2$$

- $$\frac{2\left(b^3+8a^2d-4abc\right)}{\left(b-4a\rho\right)^3}-1<-1\Leftrightarrow\frac{b^3+8a^2d-4abc}{\left(b-4a\rho\right)^3}<0\Leftrightarrow M<-1.$$

In the summary above, we have analyzed all cases that relate to the values of M, N_k and T_k affecting the number of roots to the quartic equation. We have presented a closed-form of the quartic formula corresponding to each case expressed in terms of trigonometric or hyperbolic functions. Our objective for conducting this analysis is to demonstrate how to derive solution formulas corresponding to each value of M, N_k and T_k using the general quartic formula (84b), (85b), and (93) for determining real roots with ease while minimizing computational power. Although the quartic formula may seem lengthy and impractical in application, its thorough analysis may serve as an useful tool that enables a deeper understanding of the properties of quartic equations, some of which may go unnoticed by the traditional method attributed to Lodovico Ferrari. In addition, we see that formula (99) demonstrates that a more elegant solution can be obtained by converting these formulas into radical forms.

Example. Use quartic formula indicated in *Summary of Solution to Quartic Equation in Radical Form* to find the roots of the quartic equation $x^4+x^3-x^2+3x+5=0$.

Solution. To find solution to the given equation, we do the following steps:
1. The coefficients of the given equation are a = 1, b = 1, c = -1, d = 3, and e = 5.
2. Determine the values of P, Q, R, D and S:

$$P=4ac^2-16a^2e-b^2c-2abd=4(1)(-1)^2-16(1^2)(5)-(1^2)(-1)-2(1)(1)(3)=-81$$

$$Q=3\left(b^3+8a^2d-4abc\right)=3\left(1^3\right)+8\left(1^2\right)(3)-4(1)(1)(-1)=87$$

$$R=b^2d+8abe-4acd=\left(1^2\right)(3)+8(1)(1)(5)-4(1)(-1)(3)=55$$

$$D=P^2-QR=(-81)^2-(87)(55)=1776$$

$$S=ad^2-b^2e=(1)\left(3^2\right)-\left(1^2\right)(5)=4$$

3. Determine the value of V:

$$V=\frac{3QPR-2P^3-3Q^2S}{2(D)^{3/2}}=\frac{3(87)(-81)(55)-2(-81)^{31}-3(87^2)(4)}{2\left(1776^{3/2}\right)}=-\frac{21189\sqrt{111}}{175232}\approx-1.2739...$$

4. Determine the value of ρ:
Since D>0 and V = -1.2739... < -1, we apply the formula:

135

$$\rho = -\frac{P}{Q} - \frac{2\sqrt{D}}{Q}\cosh\left(\frac{1}{3}\cosh^{-1}(|V|)\right) = \frac{27}{29} - \frac{8\sqrt{111}}{87}\cosh\left(\frac{1}{3}\cosh^{-1}\left(\frac{21189\sqrt{111}}{175232}\right)\right) \approx -0.066133\dots\Bigg)$$

Or using (99k) to get the radical form of ρ:

$$\rho = -\frac{P}{Q} - \frac{\sqrt{D}}{2Q}\left[\left(V+\sqrt{V^2-1}\right)^{1/3} + \left(V+\sqrt{V^2-1}\right)^{-1/3}\right]$$

$$= \frac{81}{87} - \frac{\sqrt{1776}}{87}\left[\left(\frac{21189\sqrt{111}}{175232} + \sqrt{\frac{517022411}{829898752}}\right)^{1/3} + \left(\frac{21189\sqrt{111}}{175232} + \sqrt{\frac{517022411}{829898752}}\right)^{-1/3}\right]$$

$$= \frac{81}{87} - \frac{\sqrt{3}}{\sqrt[3]{2}\sqrt[6]{37}}\left[\left(21189\sqrt{111} + 841\sqrt{27047}\right)^{1/3} + \left(21189\sqrt{111} + 841\sqrt{27047}\right)^{-1/3}\right]$$

5. Determine the values of O, M, and N:

$$O = b^3 + 8a^2d - 4abc = 1^3 + 8(1^2)(3) - 4(1)(1)(-1) = 29$$

$$M = b - 4a\rho = 1 - 4\rho \approx 1.2645\dots$$

$$N = 16a^2(3b\rho^2 + d - 4a\rho^3 - 2c\rho) = 16(3\rho^2 + 3 - 4\rho^3 + 2\rho) = -64\rho^3 + 48\rho^2 + 32\rho + 48 \approx 46.1122\dots$$

Then by substituting these determined values in formulas (99d)-(99g), we obtain the roots,

$$x_1 = -\rho - \frac{M^{3/2} - \sqrt{O} - \sqrt{\left(M^{3/2} - \sqrt{O}\right)^2 - N}}{4a\sqrt{M}} \approx 0.94721\dots - i\,1.2258\dots$$

$$x_2 = -\rho - \frac{M^{3/2} + \sqrt{O} - \sqrt{\left(M^{3/2} + \sqrt{O}\right)^2 - N}}{4a\sqrt{M}} \approx -1.5527\dots$$

$$x_3 = -\rho - \frac{M^{3/2} - \sqrt{O} + \sqrt{\left(M^{3/2} - \sqrt{O}\right)^2 - N}}{4a\sqrt{M}} \approx 0.94721\dots + i\,1.2258\dots\ ,$$

and

$$x_4 = -\rho - \frac{M^{3/2} + \sqrt{O} + \sqrt{\left(M^{3/2} + \sqrt{O}\right)^2 - N}}{4a\sqrt{M}} \approx -1.3417\dots\ .$$

Substituting the values of x_1, x_2, and x_3 into the given equation, we use a calculator to verify that each substitution yields zero, confirming that they are indeed the roots to the equation.

Section IV. Solving Some Higher Degree Equations with Specific Forms

In this section, we employ the same approach as in the previous sections to solve higher order equations with specific forms. Since there exist various identities involving with trigonometric or hyperbolic trigonometric functions, we only demonstrate the use of typical ones, each of which can be utilized to illustrate the remarkable aspects of this approach, which can derive multiple formulas for solving the equations.

A. Solution to Special Form of the Quintic Equation (Type 1)

In this section, we will derive a formula for a special quintic equation in the form as shown below.

$$ax^5+bx^4+cx^3+dx^2+\frac{10ad^2-4bcd+c^3}{4(5ac-2b^2)}x+\frac{10ac^3d+50a^2d^3+2b^2c^2d+4b^3d^2-40abcd^2-bc^4}{20(5ac-2b^2)^2}=0 \quad (a\neq0) \quad (200)$$

where a, b, c and d are real numbers.

We start with the quintic angle identity for tangent [2-A-9] which is expressed as

$$\tan(5x)=\frac{\tan^5 x-10\tan^3 x+5\tan x}{5\tan^4 x-10\tan^2 x+1}.$$

To derive a formula for equation (200), we do in the following steps:

1. By replacing x with $\tan^{-1}(x)$, we get:

$$\tan(5\tan^{-1}(x))=\frac{x^5-10x^3+5x}{5x^4-10x^2+1}$$

2. Multiplying both sides by a and setting each side to b gives a system of two simultaneous equations:

$$\begin{cases} \dfrac{a(x^5-10x^3+5x)}{5x^4-10x^2+1}=b & (201a) \\ a\tan(5\tan^{-1}(x))=b & (201b) \end{cases} \quad (201)$$

3. From (201a), we deduce the quintic equation in x. From (201b), we solve for x, which yields:

$$\begin{cases} a\,x^5 - 5\,b\,x^4 - 10\,a\,x^3 + 10\,b\,x^2 + 5\,a\,x - b = 0 & (202\,a) \\ x = \tan\left(\dfrac{1}{5}\tan^{-1}\left(\dfrac{b}{a}\right) + \dfrac{\pi\,k}{5}\right) & (202\,b) \end{cases} \qquad (202)$$

4. By replacing x with x/c into (202a) and (202b), then multiplying through both sides of (202a) by c^5, which gives

$$\begin{cases} a\,x^5 - 5\,b\,c\,x^4 - 10\,a\,c^2\,x^3 + 10\,b\,c^3\,x^2 + 5\,a\,c^4\,x - b\,c^5 = 0 & (203\,a) \\ x = c\tan\left[\dfrac{1}{5}\tan^{-1}\left(\dfrac{b}{a}\right) + \dfrac{\pi\,k}{5}\right] & (203\,b) \end{cases} \qquad (203)$$

5. Replacing b with b/(5c), we get:

$$\begin{cases} a\,x^5 - b\,x^4 - 10\,a\,c^2\,x^3 + 2\,b\,c^2\,x^2 + 5\,a\,c^4\,x - \dfrac{1}{5}b\,c^4 = 0 & (204\,a) \\ x = c\tan\left[\dfrac{1}{5}\tan^{-1}\left(\dfrac{b}{5\,a\,c}\right) + \dfrac{\pi\,k}{5}\right] & (204\,b) \end{cases} \qquad (204)$$

6. Replacing c with $\sqrt{c/(10\,a)}$ gives

$$\begin{cases} a\,x^5 - b\,x^4 - c\,x^3 + \dfrac{b\,c}{5\,a}x^2 + \dfrac{c^2}{20\,a}x - \dfrac{b\,c^2}{500\,a^2} = 0 & (205\,a) \\ x = \sqrt{\dfrac{c}{10\,a}}\,\tan\left[\dfrac{1}{5}\tan^{-1}\left(\dfrac{\sqrt{10\,b}}{5\sqrt{a\,c}}\right) + \dfrac{\pi\,k}{5}\right] & (205\,b) \end{cases} \qquad (205)$$

Currently, a new coefficient d can be generated for the x^2 term by substituting x with x+b/(5a) into (205a), resulting in the disappearance of coefficient b. Subsequently, we substitute x with x-b/(5a) to restore the coefficient b once the coefficient d has been determined.

7. Replacing x with x+b/(5a) gives

$$\begin{cases} a\,x^5 - \left(c + \dfrac{2\,b^2}{5\,a}\right)x^3 - \left(\dfrac{4\,b^3}{25\,a^2} + \dfrac{2\,b\,c}{5\,a}\right)x^2 + \left(\dfrac{c^2}{20\,a} - \dfrac{3\,b^4}{125\,a^3} - \dfrac{b^2\,c}{25\,a^2}\right)x + \dfrac{b\,c^2}{125\,a^2} - \dfrac{4\,b^5}{3125\,a^4} = 0 & (206\,a) \\ x = -\dfrac{b}{5\,a} + \sqrt{\dfrac{c}{10\,a}}\,\tan\left[\dfrac{1}{5}\tan^{-1}\left(\dfrac{\sqrt{10\,b}}{5\sqrt{a\,c}}\right) + \dfrac{\pi\,k}{5}\right] & (206\,b) \end{cases} \qquad (206)$$

138

8. Replacing c with (-c − 2b²/(5a)) gives

$$a x^5 + c x^3 + \frac{2bc}{5a} x^2 + \left(\frac{c^2}{20a} + \frac{2b^2 c}{25 a^2} \right) x + \frac{bc^2}{125 a^2} + \frac{4 c b^3}{625 a^3} = 0 \qquad (207\,a)$$

$$x = -\frac{b}{5a} + \frac{1}{5a} \sqrt{\frac{-2b^2 - 5ac}{2}} \tan\left[\frac{1}{5} \tan^{-1}\left(\frac{\sqrt{2}\,b}{\sqrt{-2b^2 - 5ac}} \right) + \frac{\pi k}{5} \right] \qquad (207\,b)$$

$$(207)$$

9. By setting 2bc/(5a)=d, it follows that b=5ad/(2c). Substituting d in (207) yields:

$$a x^5 + c x^3 + d x^2 + \left(\frac{c^2}{20a} + \frac{d^2}{2c} \right) x + \frac{d^3}{10 c^2} + \frac{cd}{50a} = 0 \qquad (208\,a)$$

$$x = -\frac{d}{2c} + \frac{1}{10ac} \sqrt{-10 a c^3 - 25 a^2 d^2} \tan\left[\frac{1}{5} \tan^{-1}\left(\frac{5ad}{\sqrt{-10 a c^3 - 25 a^2 d^2}} \right) + \frac{\pi k}{5} \right] \qquad (208\,b)$$

$$(208)$$

10. By applying tan(ix) = itanh(x) to (208b), which gives

$$a x^5 + c x^3 + d x^2 + \left(\frac{c^2}{20a} + \frac{d^2}{2c} \right) x + \frac{d^3}{10 c^2} + \frac{cd}{50a} = 0 \qquad (210\,a)$$

$$x = -\frac{d}{2c} + \frac{1}{10ac} \sqrt{10 a c^3 + 25 a^2 d^2} \tanh\left[\frac{1}{5} \tanh^{-1}\left(\frac{5ad}{\sqrt{10 a c^3 + 25 a^2 d^2}} \right) + \frac{\pi k}{5} \right] \qquad (210\,b)$$

$$(210)$$

11. Now, to restore b as mentioned earlier by replacing x + b/(5a), which gives:

$$a x^5 + b x^4 + \left(c + \frac{2b^2}{5a} \right) x^3 + \left(d + \frac{2b^3}{25 a^2} + \frac{3bc}{5a} \right) x^2 + \left(\frac{b^4}{125 a^3} + \frac{3 b^2 c}{25 a^2} + \frac{2bd}{5a} \right) x + \frac{b^5}{3125 a^4}$$

$$+ \frac{b^3 c}{125 a^3} + \frac{b^2 d}{25 a^2} + \frac{cd}{50a} + \frac{d^3}{10 c^2} = 0 \qquad (211\,a)$$

$$x = -\frac{b}{5a} - \frac{d}{2c} + \frac{1}{10ac} \sqrt{10 a c^3 + 25 a^2 d^2} \tanh\left(\frac{1}{5} \tanh^{-1}\left(\frac{5ad}{\sqrt{10 a c^3 + 25 a^2 d^2}} \right) \right) \qquad (211\,b)$$

$$(211)$$

12. Replacing c with c −2b²/(5a) yields

$$ax^5+bx^4+cx^3+\left(d+\frac{3bc}{5a}-\frac{4b^3}{25a^2}\right)x^2+\left(\frac{3b^2c}{25a^2}+\frac{2bd}{5a}-\frac{b^4}{125a^3}\right)x+\frac{b^3c}{125a^3}$$

$$+\frac{cd}{50a}-\frac{9b^5}{3125a^4}+\frac{4b^2d}{125a^2}+\frac{5a^2d^3}{2(5ac-2b^2)^2}=0 \qquad (212a)$$

$$x=-\frac{b}{5a}-\frac{5ad}{2(5ac-2b^2)}+\frac{\sqrt{2(5ac-2b^2)^3+625a^4d^2}}{10a(5ac-2b^2)}\tanh\left(\frac{1}{5}\tanh^{-1}\left(\frac{25a^2d}{\sqrt{2(5ac-2b^2)^3+625a^4d^2}}\right)\right) \qquad (212b)$$

$$(212)$$

13. Ultimately, after substituting d with $d-3bc/(5a)+4b^3/(25a^2)$ and performing some deductions on (213a) and (213b), we derive the solution to equation (200) in either in the form,

$$ax^5+bx^4+cx^3+dx^2+\frac{10ad^2-4abcd+c^3}{4(5ac-2b^3)}x+\frac{10ac^3d+50a^2d^3+2b^2c^2d+4b^3d^2-40abcd^2-bc^4}{20(5ac-2b^2)^2}=0 \qquad (213a)$$

$$x=\frac{bc-5ad}{2(5ac-2b^2)}+\frac{\sqrt{2(5ac-2b^2)^3+(25a^2d+4b^3-15abc)^2}}{10a(5ac-2b^2)}\tanh\left(\frac{1}{5}\tanh^{-1}\left(\frac{25a^2d+4b^3-15abc}{\sqrt{2(5ac-2b^2)^3+(25a^2d+4b^3-15abc)^2}}\right)\right) \qquad (213b)$$

$$(213)$$

or in the form, with the expressions inside radical expanded and simplified as demonstrated below:

$$ax^5+bx^4+cx^3+dx^2+\frac{10ad^2-4abcd+c^3}{4(5ac-2b^3)}x+\frac{10ac^3d+50a^2d^3+2b^2c^2d+4b^3d^2-40abcd^2-bc^4}{20(5ac-2b^2)^2}=0 \qquad (214a)$$

$$x=\frac{bc-5ad}{2(5ac-2b^2)}+\frac{\sqrt{8b^3d+10ac^3+25a^2d^2-3b^2c^2-30abcd}}{2(5ac-2b^2)}\tanh\left(\frac{1}{5}\tanh^{-1}\left(\frac{25a^2d+4b^3-15abc}{5a\sqrt{8b^3d+10ac^3+25a^2d^2-3b^2c^2-30abcd}}\right)\right) \qquad (214b)$$

$$(214)$$

Currently, it is not possible to generate a different coefficient e for the x term or f for the constant term using any methods that are presented in the prior sections. For example, if we apply the same approach as in previous sections which involves substituting a with a/d, b with b/d, and c with c/d in equation (214a) to obtain a new coefficient e, we will ultimately arrive at the same result as the original equation (214a). Similarly, if we attempt to replace x with x + e in (214a) to generate a new coefficient e, we will still end up with the same outcome as the original equation (214a) after substituting the new values of b, c, and d, respectively.

The following is a summary of solution to the special quintic equation (type 1), along with some interesting properties of this type.

Summary of Solution to Special Quintic Equation (Type 1)

Equation:
$$ax^5+bx^4+cx^3+dx^2+\frac{10ad^2-4bcd+c^3}{4(5ac-2b^2)}x+\frac{10ac^3d+50a^2d^3+2b^2c^2d+4b^3d^2-40abcd^2-bc^4}{20(5ac-2b^2)^2}=0 \quad (a\neq0)$$

Define

$$D=25\,a\left(8\,b^3\,d+10\,a\,c^3+25\,a^2\,d^2-3\,b^2\,c^2-30\,a\,b\,c\,d\right) \tag{215}$$

$$T=\frac{25\,a^2\,d+4\,b^3-15\,a\,b\,c}{\sqrt{D}} \tag{216}$$

1. Case D > 0 and (5ac-2b²) ≠ 0:

The equation has five distinct real roots:

$$x_{k+1}=\frac{bc-5ad}{2(5ac-2b^2)}+\frac{\sqrt{D}}{10\,a(5ac-2b^2)}\tan\left(\frac{1}{5}\tan^{-1}(T)+\frac{2\pi k}{5}\right)\quad k=0,1,2,3,4 \tag{217}$$

By applying (46g), the formula (217) can be expressed in the radical form,

$$x_{k+1}=\frac{bc-5ad}{2(5ac-2b^2)}+\frac{\sqrt{D}}{i\,10\,a(5ac-2b^2)}\frac{(1+iT)^{1/5}e^{2i\pi k/5}-(1-iT)^{1/5}}{(1+iT)^{1/5}e^{2i\pi k/5}+(1-iT)^{1/5}},k=0,1,2,3,4. \tag{218}$$

2. Case D < 0 and (5ac-2b²) ≠ 0:

In this case D is rewritten as

$$T=\frac{25\,a^2\,d+4\,b^3-15\,a\,b\,c}{\sqrt{|D|}}.$$

a. If -1<T<1, the equation has one real root and two pairs complex conjugate roots, provided by formula (214b):

$$x_{k+1}=\frac{bc-5ad}{2(5ac-2b^2)}+\frac{\sqrt{|D|}}{10\,a(5ac-2b^2)}\tanh\left(\frac{1}{5}\tanh^{-1}(T)+\frac{i\,2\pi k}{5}\right)\quad k=0,1,2,3,4. \tag{219}$$

By applying (46c), the formula can be expressed in the radical form,

$$x_{k+1}=\frac{bc-5ad}{2(5ac-2b^2)}+\frac{\sqrt{|D|}}{10\,a(5ac-2b^2)}\frac{(1+T)^{1/5}e^{2i\pi k/5}-(1-T)^{1/5}}{(1+T)^{1/5}e^{2i\pi k/5}+(1-T)^{1/5}},\quad k=0,1,2,3,4. \tag{220}$$

b. If T < -1 or T > 1, the equation has one root and two pairs of complex conjugate roots via formula (218):

$$x_{k+1}=\frac{bc-5ad}{2(5ac-2b^2)}+\frac{\sqrt{|D|}}{10a(5ac-2b^2)}\coth\left(\frac{1}{5}\coth^{-1}(T)+\frac{i2\pi k}{5}\right)\quad k=0,1,2,3,4. \tag{221}$$

By applying (46d), the formula can be expressed in the radical form,

$$x_{k+1}=\frac{bc-5ad}{2(5ac-2b^2)}+\frac{\sqrt{|D|}}{10a(5ac-2b^2)}\frac{(T+1)^{1/5}e^{2i\pi k/5}+(T-1)^{1/5}}{(T+1)^{1/5}e^{2i\pi k/5}-(T-1)^{1/5}},\quad k=0,1,2. \tag{222}$$

3. Case D = 0:

a. 5ac-2b² = 0

By (218), the equation has a repeated root:

$$x_1=x_2=x_3=x_4=x_5=\frac{5ac-6b^2}{20ab}. \tag{223}$$

b. 5ac-2b² ≠ 0:

The equation has five simple roots, four of which are repeated:

$$\begin{cases} x_{1,3,4,5}=\dfrac{bc-5ad}{2(5ac-2b^2)} & (224\,a) \\ \text{and} \\ x_2=\dfrac{2b^3+10a^2d-7abc}{a(5ac-2b^2)} & (224\,b) \end{cases} \tag{224}$$

Notice that formulas (223) and (224) hold true not only for equation (200) but also for any generic quintic equation. This is due to the fact that these formulas solely take into account the coefficients a, b, c, and d of the quintic equation, rendering them universally applicable to all quintic equations.

B. Solution to Special Form of the Quintic Equation (Type 2)

In this section, we will explore two quintic formulas, each is used to solve a special quintic equation with distinct coefficient structures in the forms illustrated below.

$$\begin{cases} ax^5+\dfrac{b^2}{5c}x^4+bx^2+c=0 \quad (a\neq0,c\neq0) & (225\,a) \\ ax^5+bx^4+cx^2+\dfrac{c^2}{5b}=0. \quad (a\neq0,b\neq0) & (225\,b) \end{cases} \tag{225}$$

142

To derive the quintic formulas for (225a) and (225b), we use formula (46*b) with n = 1/5 and k = 0. Then by replacing x with sinh^{-1}(x), we get:

$$\sinh\left(5\sinh^{-1}(x)\right) = 16\,x^5 + 20\,x^3 + 5\,x \qquad (255\,c)$$

By replacing x with x/2 and multiplying both sides by 2, which gives

$$2\sinh\left(5\sinh^{-1}\left(\frac{x}{2}\right)\right) = x^5 + 5\,x^3 + 5\,x \qquad (225\,d)$$

We now proceed to construct the coefficients of the equation on the right-hand side of (225d) to match the coefficients of equation (225a) as shown in the following steps:

1. Let m be a real number. Subtracting m from both sides of (225d) and setting both sides to zero yields a system of two simultaneous equations:

$$\begin{cases} x^5 + 5\,x^3 + 5\,x - m = 0 & (226\,a) \\ x_{k+1} = 2\sinh\left(\frac{1}{5}\sinh^{-1}\left(\frac{m}{2}\right) + \frac{i\,2\,\pi\,k}{5}\right), k = 0,1,2,3,4. & (226\,b) \end{cases} \qquad (226)$$

2. To construct the coefficient c for the x^3 term, we replace x with x/c$^{1/2}$ for c ≥ 0 into both (226a) and (226b), then multiply through both sides of (226a) by c$^{5/2}$, and extend the multiplication across each term of its left side, which gives

$$\begin{cases} x^5 + 5\,c\,x^3 + 5\,c^2\,x - c^{5/2}\,m = 0 & (227\,a) \\ x_{k+1} = 2\,c^{1/2}\sinh\left(\frac{1}{5}\sinh^{-1}\left(\frac{m}{2}\right) + \frac{i\,2\,\pi\,k}{5}\right). & (227\,b) \end{cases} \qquad (227)$$

3. Replacing m with -m/c$^{5/2}$ and c ≠ 0 gives

$$\begin{cases} x^5 + 5\,c\,x^3 + 5\,c^2\,x + m = 0 & (228\,a) \\ x_{k+1} = -2\,c^{1/2}\sinh\left(\frac{1}{5}\sinh^{-1}\left(\frac{m}{2\,c^{5/2}}\right) + \frac{i\,2\,\pi\,k}{5}\right), \; k = 0,1,2,3,4. & (228\,b) \end{cases} \qquad (228)$$

4. Substituting c by c/5 gives

$$\begin{cases} x^5 + c\,x^3 + \frac{1}{5}c^2\,x + m = 0 & (229\,a) \\ x_{k+1} = -2\sqrt{\frac{c}{5}}\sinh\left(\frac{1}{5}\sinh^{-1}\left(\frac{25\sqrt{5}\,m}{2\,c^{5/2}}\right) + \frac{i\,2\,\pi\,k}{5}\right), k = 0,1,2,3,4. & (229\,b) \end{cases} \qquad (229)$$

Notice that the equation (229a) is called de Moivre's Quintic [11], and its roots are given by (229b).

5. Next, we construct the coefficient a for x^5 term by replacing m with m/a for a ≠ 0 and multiplying through both sides of (229a) by a, which gives

$$
\begin{cases}
a x^5 + a c x^3 + \dfrac{1}{5} a c^2 x + m = 0 \quad (a \neq 0) & (230a) \\[4mm]
x_{k+1} = -2 \sqrt{\dfrac{c}{5}} \sinh\left(\dfrac{1}{5} \sinh^{-1}\left(\dfrac{25\sqrt{5}\,m}{2\,a\,c^{5/2}} \right) + \dfrac{i\,2\pi k}{5} \right), k = 0,1,2,3,4. & (230b)
\end{cases}
\tag{230}
$$

6. Replacing c with c/a gives

$$
\begin{cases}
a x^5 + c x^3 + \dfrac{c^2}{5 a} x + m = 0 \quad (a \neq 0) & (231a) \\[4mm]
x_{k+1} = -2 \sqrt{\dfrac{c}{5 a}} \sinh\left(\dfrac{1}{5} \sinh^{-1}\left(\dfrac{25\sqrt{5}\,m\,a^{3/2}}{2\,c^{5/2}} \right) + \dfrac{i\,2\pi k}{5} \right), k = 0,1,2,3,4. & (231b)
\end{cases}
\tag{231}
$$

At this point, we can obtain a similar result to that of quintic equation (type 1) when trying to derive the coefficients b for the x^4 term and d for the x term. However, the coefficients are slightly different due to the use of distinct hyperbolic functions. To explore another form of the equation that has x^4 and x^2 term, we substitute x with 1/x, which yields the following expression.

$$
\begin{cases}
a + c x^2 + \dfrac{c^2}{5 a} x^4 + m x^5 = 0 \quad (a \neq 0) & (232a) \\[4mm]
x_{k+1} = -\dfrac{\sqrt{5 a}}{2\sqrt{c}\,\sinh\left(\dfrac{1}{5} \sinh^{-1}\left(\dfrac{25\sqrt{5}\,m\,a^{3/2}}{2\,c^{5/2}} \right) + \dfrac{i\,2\pi k}{5} \right)}, k = 0,1,2,3,4. & (232b)
\end{cases}
\tag{232}
$$

7. If we rename all coefficients and terms such that m is renamed as a, c is renamed as b, and a is renamed as c, we get:

$$
\begin{cases}
a x^5 + \dfrac{b^2}{5 c} x^4 + b x^2 + c = 0 \quad (a \neq 0, c \neq 0) & (233a) \\[4mm]
x_{k+1} = -\dfrac{\sqrt{5 c}}{2\sqrt{b}\,\sinh\left(\dfrac{1}{5} \sinh^{-1}\left(\dfrac{25\sqrt{5}\,a\,c^{3/2}}{2\,b^{5/2}} \right) + \dfrac{i\,2\pi k}{5} \right)}, k = 0,1,2,3,4. & (233b)
\end{cases}
\tag{233}
$$

Consistent with our method, we have completed derivation of the solution formula (233b) for equation (225a). This equation, (225a) or (233a), can be considered as an alternative depiction of de Moivre's Quintic [11]. It stands as a notable incomplete variant of the general quintic equation.

Summary of Solution to Special Quintic Equation (Type 2)

Quintic equation in the form: $a x^5 + \dfrac{b^2}{5c} x^4 + b x^2 + c = 0 \quad (a \neq 0, c \neq 0)$

Let $T = \dfrac{25\sqrt{5}\, a\, c^{3/2}}{2 b^{5/2}}.$ (235)

i. If c > 0 and b > 0, the equation has one real root and four complex roots:

$$x_{k+1} = -\dfrac{\sqrt{5c}}{2\sqrt{b}\sinh\left(\dfrac{1}{5}\sinh^{-1}(T) + \dfrac{i 2\pi k}{5}\right)}.$$ (236)

ii. If c < 0 and b > 0, then

$$T = \dfrac{25\sqrt{5}\, a\, |c|^{3/2}}{2 b^{5/2}}.$$ (237)

- If $|T| \leq 1$, the equation has five real roots:

$$x_{k+1} = \dfrac{\sqrt{5|c|}}{2\sqrt{b}\sin\left(\dfrac{1}{5}\sin^{-1}(T) + \dfrac{2\pi k}{5}\right)}.$$ (236)

- If $T \geq 1$, the equation has one real root and four complex roots:

$$x_{k+1} = \dfrac{\sqrt{5|c|}}{2\sqrt{b}\cosh\left(\dfrac{1}{5}\cosh^{-1}(T) + \dfrac{i 2\pi k}{5}\right)}.$$ (237)

- If $T \leq -1$, the equation has one real root and four complex roots:

$$x_{k+1} = -\dfrac{\sqrt{5|c|}}{2\sqrt{b}\cosh\left(\dfrac{1}{5}\cosh^{-1}(|T|) + \dfrac{i 2\pi k}{5}\right)}.$$ (238)

iii. If c < 0 and b < 0, then

$$T = \frac{25\sqrt{5}\,a\,|c|^{3/2}}{2\,|b|^{5/2}}.$$

(239)

The equation has one real root and four complex roots:

$$x_{k+1} = \frac{\sqrt{5}\,|c|}{2\sqrt{|b|}\,\sinh\left(\frac{1}{5}\sinh^{-1}(T) + \frac{i\,2\pi k}{5}\right)}.$$

(240)

iv. If c ≥ 0 and b < 0, then

$$T = \frac{25\sqrt{5}\,a\,c^{3/2}}{2\,|b|^{5/2}}.$$

(241)

- If |T| ≤ 1, the equation has one real root and four complex roots:

$$x_{k+1} = -\frac{\sqrt{5}\,c}{2\sqrt{|b|}\,\sin\left(\frac{1}{5}\sin^{-1}(T) + \frac{2\pi k}{5}\right)}.$$

(242)

- If |T| ≥ 1, the equation has one real root and four complex roots:

$$\begin{cases} x_{k+1} = -\dfrac{\sqrt{5}\,c}{2\sqrt{|b|}\,\cosh\left(\dfrac{1}{5}\cosh^{-1}(T) + \dfrac{i\,2\pi k}{5}\right)}, & T \geq 1 \\[4mm] x_{k+1} = \dfrac{\sqrt{5}\,c}{2\sqrt{|b|}\,\cosh\left(\dfrac{1}{5}\cosh^{-1}(T) + \dfrac{i\,2\pi k}{5}\right)}, & T \leq -1. \end{cases}$$

(243)

(244)

We continue to proceed the transformation process to derive a solution formula for (225b) by the following steps:

8. Replacing b with $\sqrt{5bc}$ into (233) gives:

$$\begin{cases} a\,x^5 + b\,x^4 + \sqrt{5\,b\,c}\,x^2 + c = 0 \quad (a \neq 0) & (245a) \\[4mm] x_{k+1} = -\dfrac{\sqrt[4]{5}\,c}{2\sqrt[4]{b}\,\sinh\left(\dfrac{1}{5}\sinh^{-1}\left(\dfrac{5\,a\sqrt[4]{5}\,c}{2\,b^{5/4}}\right) + \dfrac{i\,2\pi k}{5}\right)}, & k = 0,1,2,3,4. \quad (245b) \end{cases}$$

(245)

9. Replacing c with $c^2/(5b)$ gives the form of (225b), namely

$$\begin{cases} a\,x^5 + b\,x^4 + c\,x^2 + \dfrac{c^2}{5\,b} = 0 \quad (a \neq 0, b \neq 0) & (246\,a) \\[4mm] x_{k+1} = -\dfrac{\sqrt{c}}{2\sqrt{b}\,\sinh\!\left(\dfrac{1}{5}\,\sinh^{-1}\!\left(\dfrac{5\,a\,\sqrt{c}}{2\,b^{3/2}}\right) + \dfrac{i\,2\,\pi\,k}{5}\right)} \quad, k = 0,1,2,3,4. & (246\,b) \end{cases} \qquad (246)$$

Due to the similarity between formulas (246) and (233), we're skipping the summary of (246) here and leaving it for readers. Any updates, including more comprehensive details from different chapters, will be incorporated into the republished edition or shared on an authorized website, subject to approval.

Remarks

1. The proof of (214) demonstrates the outcome in the form of a hyperbolic tangent, yielding results that are applicable to other tangent functions and hyperbolic cotangent functions due to the presence of the imaginary unit. Interestingly, this pattern holds true for all polynomials when their highest power is an odd degree.

2. By utilizing the equality (255c) for hyperbolic sine, we obtain the derivation of (246), which is relevant to hyperbolic cosine and its inverse in Summary section. This implies that if we employ the quadruple angle equality for hyperbolic cosine as the initial premise, the outcome would be slightly different in terms of domain, but fundamentally similar.

3. In terms of the solving method, each step outlined in the solution process yields roots to a specific equation. For instance, completing five steps results in solutions for five slightly different forms of equations. This flexibility is a remarkable feature of this method, achieved through a straightforward single substitution approach at each step.

4. Through this method, equations (225a), (225b) and (245a) are considered to belong to a specific quintic family, characterized by varying coefficient structures.

C. Solution to Special Form of the Sextic Equation

In this section, we expand upon the solution obtained for the cubic equation mentioned in (42) and utilize the result to deduce the solution for a unique sextic equation presented in the form,

$$a\,x^6 + b\,x^5 + c\,x^4 + d\,x^3 + e\,x^2 + \frac{b\,e^2}{c^2}\,x + \frac{a\,e^3}{c^3} = 0 \quad (a \neq 0, c \neq 0). \qquad (247)$$

We derive the solution to equation (247) in the following steps:

147

1. Define

$$\beta_{k+1}=-\frac{b}{3a}+\frac{2}{3a}\sqrt{b^2-3ac}\cdot\cosh\left(\frac{1}{3}\cosh^{-1}\left(\frac{9abc-2b^3-27a^2d}{2(b^2-3ac)^{3/2}}\right)+\frac{i2\pi k}{3}\right),k=0,1,2. \qquad (248)$$

2. By replacing x by x + e/x into (42a) and (42b), then multiplying through both sides of (42a) by x^3 and solving for x from (42b), which gives

$$\begin{cases} ax^6+bx^5+(3ae+c)x^4+(2be+d)x^3+(3ae^2+cg^3e)x^2+be^2x+ae^3=0 & (249a)\\ x+\dfrac{e}{x}=\beta_{k+1}\Rightarrow x_{1...6}=\dfrac{\beta_{k+1}}{2}\pm\dfrac{1}{2}\sqrt{\beta^2_{k+1}-4e} \quad \text{for } k=0,1,2. & (249b) \end{cases} \qquad (249)$$

Formula (249b) can be understood in this way: For each value of k, we have has two roots. Hence, there are six roots in total.

3. Substituting c by (c - 3ah) gives:

$$\begin{cases} ax^6+bx^5+cx^4+(d+2be)x^3+cex^2+be^2x+ae^3=0 \quad (a\neq0) & (250a)\\[2mm] x_{1...6}=\dfrac{\beta_{k+1}}{2}\pm\dfrac{1}{2}\sqrt{\beta^2_{k+1}-4e} & (250b)\\[2mm] \text{where} & (250)\\[2mm] \beta_{k+1}=-\dfrac{b}{3a}+\dfrac{2}{3a}\sqrt{9a^2e+b^2-3ac}\cdot\cosh\left(\dfrac{1}{3}\cosh^{-1}\left(\dfrac{9abc-2b^3-27a^2be-27a^2d}{2(9a^2e+b^2-3ac)^{3/2}}\right)\right) & (250c)\\[2mm] \text{for } k=0,1,2. \end{cases}$$

4. Replacing d with (d - 2be) gives

$$\begin{cases} ax^6+bx^5+cx^4+dx^3+cex^2+be^2x+ae^3=0 \quad (a\neq0) & (251a)\\[2mm] x_{1...6}=\dfrac{\beta_{k+1}}{2}\pm\dfrac{1}{2}\sqrt{\beta^2_{k+1}-4e} & (251b)\\[2mm] \text{where} & (251)\\[2mm] \beta_{k+1}=-\dfrac{b}{3a}+\dfrac{2}{3a}\sqrt{9a^2e+b^2-3ac}\cdot\cosh\left(\dfrac{1}{3}\cosh^{-1}\left(\dfrac{27a^2be+9abc-2b^3-27a^2d}{2(9a^2e+b^2-3ac)^{3/2}}\right)\right) & (251c)\\[2mm] \text{for } k=0,1,2. \end{cases}$$

5. By replacing e with e/c, which gives the solution to sextic equation (247)

148

$$ax^6+bx^5+cx^4+dx^3+ex^2+\frac{be^2}{c^2}x+\frac{ae^3}{c^3}=0 \qquad (a\neq0,c\neq0) \tag{252a}$$

$$x_{1\ldots6}=\frac{\beta_{k+1}}{2}\pm\frac{1}{2}\sqrt{\beta^2_{k+1}-\frac{4e}{c}} \tag{252b}$$

where

$$\beta_{k+1}=-\frac{b}{3a}+\frac{2}{3a}\sqrt{\frac{9a^2e+b^2c-3ac^2}{c}}\cdot\cosh\left(\frac{1}{3}\cosh^{-1}\left(\frac{9abc^{5/2}+27a^2b\sqrt{c}e-2b^3c^{3/2}-27a^2c^{3/2}d}{2\left(9a^2e+b^2c-3ac^2\right)^{3/2}}\right)\right) \tag{252c}$$

for $k=0,1,2.$

$$(252)$$

In accordance with sextic formula (252b) to sextic equation (252a), we can obtain six roots in total. For each value of k, we have two roots. Specifically, when k = 0, the roots are x_1 and x_2; when k = 1, the roots are x_3 and x_4; and when k = 2, the roots are x_5 and x_6. Notice that the choice of defining β in (248), which can use either hyperbolic cosine (42b) or trigonometric cosine (40b), does not affect the final results.

Summary of Solution to Special Sextic Equation

Equation:
$$a x^6+b x^5+c x^4+d x^3+e x^2+\frac{b e^2}{c^2}x+\frac{a e^3}{c^3}=0 \quad (a\neq0,c\neq0)$$

Define

$$D=9a^2e+b^2c-3ac^2 \tag{253}$$

$$T=\frac{9abc^{5/2}-2b^3c^{3/2}+27a^2b\sqrt{c}e-27a^2c^{3/2}d}{2D^{3/2}} \tag{254}$$

+> Case D > 0:

If T ≥ 1, the equation has two real roots and four complex conjugate roots,

$$x_{1,\ldots,6}=\frac{\beta_{k+1}}{2}\pm\frac{1}{2}\sqrt{\beta^2_{k+1}-\frac{4e}{c}}. \tag{255}$$

where

$$\beta_{k+1}=-\frac{b}{3a}+\frac{2}{3a}\sqrt{\frac{D}{c}}\cdot\cosh\left(\frac{1}{3}\cosh^{-1}(T)+\frac{i2\pi k}{3}\right),\ k=0,1,2. \tag{256}$$

If T ≤ -1, the equation has two real roots and four complex conjugate roots,

$$x_{1,\dots,6} = \frac{\beta_{k+1}}{2} \pm \frac{1}{2}\sqrt{\beta_{k+1}^2 - \frac{4e}{c}}.$$ (257)

where

$$\beta_{k+1} = -\frac{b}{3a} + \frac{2}{3a}\sqrt{\frac{D}{c}} \cdot \cosh\left(\frac{1}{3}\cosh^{-1}(|T|) + \frac{\pi i(2k+1)}{3}\right), \quad k = 0,1,2.$$ (258)

If -1≤T≤1, the equation has six distinct real roots

$$x_{1,\dots,6} = \frac{\beta_{k+1}}{2} \pm \frac{1}{2}\sqrt{\beta_{k+1}^2 - \frac{4e}{c}}. \quad \text{when} \quad \beta_{k+1}^2 - \frac{4e}{c} \geq 0,$$ (259)

or the equation has three pair complex conjugate roots,

$$x_{1,\dots,6} = \frac{\beta_{k+1}}{2} \pm \frac{i}{2}\sqrt{\left|\beta_{k+1}^2 - \frac{4e}{c}\right|} \quad \text{when} \quad \beta_{k+1}^2 - \frac{4e}{c} < 0,$$ (260)

where

$$\beta_{k+1} = -\frac{b}{3a} + \frac{2}{3a}\sqrt{\frac{D}{c}} \cdot \cos\left(\frac{1}{3}\cos^{-1}(T) + \frac{2\pi k}{3}\right), \quad k = 0,1,2.$$ (261)

+> *Case D < 0:*

In this case, T becomes

$$T = \frac{9abc^{5/2} - 2b^3c^{3/2} + 27a^2b\sqrt{c}e - 27a^2c^{3/2}d}{2|D|^{3/2}}.$$ (262)

If $\beta_{k+1}^2 - \frac{4e}{c} \geq 0$, the equation has two real roots and four complex conjugate roots,

$$x_{1,\dots,6} = \frac{\beta_{k+1}}{2} \pm \frac{1}{2}\sqrt{\beta_{k+1}^2 - \frac{4e}{c}}.$$ (263)

If $\beta_{k+1}^2 - \frac{4e}{c} < 0$, the equation has three pair complex conjugate roots,

$$x_{1,\dots,6}=\frac{\beta_{k+1}}{2}\pm\frac{i}{2}\sqrt{\left|\beta_{k+1}^2-\frac{4e}{c}\right|}.$$

(264)

where

$$\beta_{k+1}=-\frac{b}{3a}-\frac{2}{3a}\sqrt{\frac{|D|}{c}}\cdot\sinh\left(\frac{1}{3}\sinh^{-1}(T)+\frac{\pi i(2k-1)}{3}\right),\ k=0,1,2.$$

(265)

If -1 ≤ T ≤ 1, the equation has two real roots and four complex conjugate roots,

$$x_{1,\dots,6}=\frac{\beta_{k+1}}{2}\pm\frac{1}{2}\sqrt{\beta_{k+1}^2-\frac{4e}{c}}\quad\text{when}\quad\beta_{k+1}^2-\frac{4e}{c}\geq0,$$

(266)

or the equation has six complex conjugate roots,

$$x_{1,\dots,6}=\frac{\beta_{k+1}}{2}\pm\frac{1}{2}\sqrt{\beta_{k+1}^2-\frac{4e}{c}}\quad\text{when}\quad\beta_{k+1}^2-\frac{4e}{c}<0.$$

(267)

+> Case D = 0:
The equation has six roots, and the presence of complex roots depends on the signs of the coefficient c, the square root in (268), and the value of k = 1, 2 in (269) as shown below:

$$x_{1,\dots,6}=\frac{\beta_{k+1}}{2}\pm\frac{1}{2}\sqrt{\beta_{k+1}^2-\frac{4e}{c}},$$

(268)

where

$$\beta_{k+1}=-\frac{b}{3a}+\frac{1}{3a\sqrt{c}}\sqrt[3]{9abc^{5/2}-2b^3c^{3/2}+27a^2b\sqrt{c}e-27a^2c^{3/2}d}\ e^{i2\pi k/3},k=0,1,2.$$

(269)

Note:
1. To proof for formula (269), we use the result of (45) in which we define:

$$A=\frac{2}{3a\sqrt{c}}\quad\text{and}\quad B=\frac{1}{2}\left(9abc^{5/2}-2b^3c^{3/2}+27a^2b\sqrt{c}e-27a^2c^{3/2}d\right).$$

2. It is not hard to see that formulas (252) not only provide roots to the sextic equation (247) but also act as a means to derive solutions for equations presented in the following forms:

$$ax^6+cx^4+dx^3+ex^2+\frac{ae^3}{c^3}=0,\quad\text{obtained by setting b = 0 from (252),}$$

and

$$a x^6 + b x^5 + c x^3 + d x + \frac{a d^{3/2}}{b^{3/2}} = 0, \quad \text{derived by substituting x with 1/x in (252).}$$

D. Solution to Special Form of the Septic Equation

In this section, we will derive the solution to special septic equation in the forms,

$$y^7 + 7 c y^5 + 14 c^2 y^3 + 7 c^3 y + m = 0$$

and

$$a x^7 + b x^6 + c x^4 + \frac{c^2}{4 b} x^2 + \frac{c^3}{56 b^2} = 0 \quad (a \neq 0, b \neq 0).$$

Similar to the process applied in the quintic equation (type 2), we can begin by using formula (46*b) with n = 1/7, k = 0. By substituting x with sinh^{-1}(x), we obtain the following equation:

$$\sinh\left(7 \sinh^{-1}(x)\right) = 64 x^7 + 112 x^5 + 56 x^3 + 7 x$$

1. By replacing x with x/2, which gives

$$2 \sinh\left(7 \sinh^{-1}\left(\frac{x}{2}\right)\right) = x^7 + 7 x^5 + 14 x^3 + 7 x \tag{271}$$

2. Let m be a real number. Subtracting m from both sides of (271) and setting both sides to zero yields a system of two simultaneous equations:

$$\begin{cases} x^7 + 7 x^5 + 14 x^3 + 7 x - m = 0 & (272\,a) \\ x_{k+1} = 2 \sinh\left(\frac{1}{7} \sinh^{-1}\left(\frac{m}{2}\right) + \frac{i 2 \pi k}{7}\right), k = 0, 1, \dots, 6. & (272\,b) \end{cases} \tag{272}$$

3. Replacing x with x/c$^{1/2}$ for c ≥ 0 into both (272a) and (272b), then multiplying through both sides of (272a) by c$^{7/2}$, which gives

$$\begin{cases} x^7 + 7 c x^5 + 14 c^2 x^3 + 7 c^3 x - c^{7/2} m = 0 & (273\,a) \\ x_{k+1} = 2 \sqrt{c} \sinh\left(\frac{1}{7} \sinh^{-1}\left(\frac{m}{2}\right) + \frac{i 2 \pi k}{7}\right), k = 0, 1, \dots, 6. & (273 b) \end{cases} \tag{273}$$

4. Replacing m with -m/c^{7/2} gives

$$\begin{cases} x^7+7\,c\,x^5+14\,c^2\,x^3+7\,c^3\,x+m=0 & (274\,a) \\[2mm] x_{k+1}=-2\sqrt{c}\,\sinh\left(\dfrac{1}{7}\sinh^{-1}\left(\dfrac{m}{2\,c^{7/2}}\right)+\dfrac{i\,2\,\pi\,k}{7}\right),\,k=0,1,\dots,6. & (274\,b) \end{cases} \qquad (274)$$

5. Substituting c by c/7 gives

$$\begin{cases} x^7+c\,x^5+\dfrac{2}{7}c^2\,x^3+\dfrac{1}{49}c^3\,x+m=0 & (275\,a) \\[2mm] x_{k+1}=-\dfrac{2\sqrt{c}}{\sqrt{7}}\,\sinh\left(\dfrac{1}{7}\sinh^{-1}\left(\dfrac{343\,m\sqrt{7}}{2\,c^{7/2}}\right)+\dfrac{i\,2\,\pi\,k}{7}\right),\,k=0,1,\dots,6. & (275\,b) \end{cases} \qquad (275)$$

6. Next, we construct the coefficient a for x^7 term by replacing m with m/a for a ≠ 0 and multiplying through both sides of (275a) by a, which gives

$$\begin{cases} a\,x^7+a\,c\,x^5+\dfrac{2}{7}a\,c^2\,x^3+\dfrac{1}{49}a\,c^3\,x+m=0 \quad (a\neq0) & (276\,a) \\[2mm] x_{k+1}=-\dfrac{2\sqrt{c}}{\sqrt{7}}\,\sinh\left(\dfrac{1}{7}\sinh^{-1}\left(\dfrac{343\,m\sqrt{7}}{2\,a\,c^{7/2}}\right)+\dfrac{i\,2\,\pi\,k}{7}\right),\,k=0,1,\dots,6. & (276\,b) \end{cases} \qquad (276)$$

7. Replacing c with c/a gives

$$\begin{cases} a\,x^7+c\,x^5+\dfrac{2\,c^2}{7\,a}x^3+\dfrac{c^3}{49\,a^2}x+m=0 \quad (a\neq0) & (277\,a) \\[2mm] x_{k+1}=-\dfrac{2\sqrt{c}}{\sqrt{7\,a}}\,\sinh\left(\dfrac{1}{7}\sinh^{-1}\left(\dfrac{343\,m\sqrt{7}\,a^{5/2}}{2\,c^{7/2}}\right)+\dfrac{i\,2\,\pi\,k}{7}\right),\,k=0,1,\dots,6. & (277\,b) \end{cases} \qquad (277)$$

8. If we substitute x with x+b/(7a), a≠0, we can determine the coefficients b for the x^6 term and d for the x^4 term. However, this approach may result in cumbersome formulas, similar to the result we presented for the quintic equation (type 1). To maintain the current simple style of this self-story, we choose a simpler presentation to depict the remarkable feature of the septic equation by replacing x with 1/x, which gives

$$\begin{cases} a+c\,x^2+\dfrac{2\,c^2}{7\,a}x^4+\dfrac{c^3}{49\,a^2}x^6+m\,x^7=0 \quad (a\neq0,\,m\neq0) & (278\,a) \\[4mm] x_{k+1}=-\dfrac{\sqrt{7\,a}}{2\sqrt{c}\,\sinh\left(\dfrac{1}{7}\sinh^{-1}\left(\dfrac{343\,m\sqrt{7}\,a^{5/2}}{2\,c^{7/2}}\right)+\dfrac{i\,2\,\pi\,k}{7}\right)},\,k=0,1,\dots,6. & (278\,b) \end{cases} \qquad (278)$$

9. If we rename all coefficients and terms such that m is renamed as a ($a \neq 0$), c is renamed as b, and a is renamed as c, we get:

$$\left\{ \begin{array}{l} a\,x^7 + \dfrac{b^3}{49\,c^2}\,x^6 + \dfrac{2\,b^2}{7\,c}\,x^4 + b\,x^2 + c = 0 \quad (a \neq 0, c \neq 0) \qquad (279\,a) \\[4mm] x_{k+1} = -\dfrac{\sqrt{7\,c}}{2\sqrt{b}\,\sinh\left(\dfrac{1}{7}\sinh^{-1}\left(\dfrac{343\,a\,\sqrt{7}\,c^{5/2}}{2\,b^{7/2}}\right) + \dfrac{i\,2\,\pi\,k}{7}\right)} \quad , k = 0,1,\ldots,6. \qquad (279\,b) \end{array} \right. \qquad (279)$$

10. Assuming $b \geq 0$, by replacing b with $\sqrt[3]{49\,b\,c^2}$, which gives

$$\left\{ \begin{array}{l} a\,x^7 + b\,x^6 + 2\sqrt[3]{7\,b^2\,c}\,x^4 + \sqrt[3]{49\,b\,c^2}\,x^2 + c = 0 \quad (a \neq 0) \qquad (280\,a) \\[4mm] x_{k+1} = -\dfrac{\sqrt{7\,c}}{2\sqrt[6]{b}\,\sinh\left(\dfrac{1}{7}\sinh^{-1}\left(\dfrac{7\,a\,\sqrt[6]{7\,c}}{2\,b\,\sqrt[6]{b}}\right) + \dfrac{i\,2\,\pi\,k}{7}\right)} \quad , k = 0,1,\ldots,6. \qquad (280\,b) \end{array} \right. \qquad (280)$$

11. Replacing c with $c^3/(56b^2)$ and $b \neq 0$ gives the solution to the desired equation,

$$\left\{ \begin{array}{l} a\,x^7 + b\,x^6 + c\,x^4 + \dfrac{c^2}{4\,b}\,x^2 + \dfrac{c^3}{56\,b^2} = 0 \quad (a \neq 0, b \neq 0) \qquad (281\,a) \\[4mm] x_{k+1} = -\dfrac{\sqrt{c}}{2\sqrt{2\,b}\,\sinh\left(\dfrac{1}{7}\sinh^{-1}\left(\dfrac{7\,a\,\sqrt{c}}{2\,b\,\sqrt{2\,b}}\right) + \dfrac{i\,2\,\pi\,k}{7}\right)} \quad , k = 0,1,\ldots,6. \qquad (281\,b) \end{array} \right. \qquad (281)$$

Summary of Solution to Special Septic Equation

Equation: $\quad a\,x^7 + b\,x^6 + c\,x^4 + \dfrac{c^2}{4\,b}\,x^2 + \dfrac{c^3}{56\,b^2} = 0 \quad (a \neq 0, b \neq 0).$

Define $\quad T = \dfrac{7\,a\,\sqrt{c}}{(2\,b)^{3/2}}$ $\qquad (282)$

i. If $c \geq 0$ and $b > 0$, the equation has one real root and six complex roots, given by (281b), which can be simplified to the form,

$$x_{k+1} = -\frac{\sqrt{c}}{2\sqrt{2b}\sinh\left(\frac{1}{7}\sinh^{-1}(T)+\frac{i2\pi k}{7}\right)}. \tag{283}$$

ii. If c < 0 and b > 0, then

$$T = \frac{7a\sqrt{|c|}}{(2b)^{3/2}} \tag{284}$$

➤ If |T| ≤ 1, the equation has six real roots:

$$x_{k+1} = \frac{\sqrt{|c|}}{2\sqrt{2b}\sin\left(\frac{1}{7}\sin^{-1}(T)+\frac{2\pi k}{7}\right)} \quad, k=0,1,\dots,6. \tag{285}$$

➤ If T ≥ 1, the equation has one real root and six complex roots:

$$x_{k+1} = \frac{\sqrt{|c|}}{2\sqrt{2b}\cosh\left(\frac{1}{7}\cosh^{-1}(T)+\frac{i2\pi k}{7}\right)} \quad, k=0,1,\dots,6. \tag{286}$$

➤ If T ≤ -1, the equation has one real root and six complex roots:

$$x_{k+1} = -\frac{\sqrt{|c|}}{2\sqrt{2b}\cosh\left(\frac{1}{7}\cosh^{-1}(|T|)+\frac{i2\pi k}{7}\right)} \quad, k=0,1,\dots,6. \tag{287}$$

iii. If c < 0 and b < 0, then

$$T = \frac{7a\sqrt{|c|}}{(2|b|)^{3/2}} \tag{288}$$

The equation has one real root and six complex roots:

$$x_{k+1} = \frac{\sqrt{|c|}}{2\sqrt{2|b|}\sinh\left(\frac{1}{7}\sinh^{-1}(T)+\frac{i2\pi k}{7}\right)} \quad, k=0,1,\dots,6. \tag{289}$$

iv. If c ≥ 0 and b < 0, then

$$T = \frac{7a\sqrt{c}}{(2|b|)^{3/2}} \tag{290}$$

➤ If |T| ≤ 1, the equation has one real root and six complex roots:

$$x_{k+1}=-\frac{\sqrt{c}}{2\sqrt{2}|b|\sin\left(\frac{1}{7}\sin^{-1}(T)+\frac{2\pi k}{7}\right)}\ ,k=0,1,\dots,6. \tag{291}$$

➤ If |T| ≥ 1, the equation has one real root and six complex roots:

$$x_{k+1}=-\frac{\sqrt{c}}{2\sqrt{2}|b|\cosh\left(\frac{1}{7}\cosh^{-1}(T)+\frac{i2\pi k}{7}\right)}\ ,T\geq1\ ,k=0,1,\dots,6. \tag{292}$$

$$x_{k+1}=\frac{\sqrt{c}}{2\sqrt{2}|b|\cosh\left(\frac{1}{7}\cosh^{-1}(|T|)+\frac{i2\pi k}{7}\right)}\ ,T\leq-1\ ,k=0,1,\dots,6. \tag{293}$$

Remarks

Readers might find equation (281a) intriguing when expressed in another form involving x^5. In this case, we present the solution without delving into the specifics of the substitution for x with x – b/(7a), a≠0. The solving process is then repeated accordingly.

Derived equation:

$$ax^7-bx^5+cx^4-\left(\frac{4\sqrt{b}c}{\sqrt{21}a}-\frac{25b^2}{63a}\right)x^3+\left(\frac{bc}{21a}-\frac{23b^{5/2}}{3(21a)^{3/2}}+\frac{\sqrt{3}c^2}{4\sqrt{7}ab}\right)x^2+\left(\frac{6b^{3/2}c}{(21a)^{3/2}}-\frac{5b^3}{441a^2}-\frac{c^2}{14a}\right)x$$

$$+\frac{\sqrt{b}c^2}{28\sqrt{21}a^{3/2}}-\frac{2b^{7/2}}{1323\sqrt{21}a^{5/2}}-\frac{5\sqrt{3}bc^2}{196\sqrt{7}a^{3/2}}+\frac{3c^3}{392ab}+\frac{19b^2c}{6174a^2}=0 \tag{281c}$$

Solution:

$$x_{k+1}=\sqrt{\frac{b}{21a}-\frac{\sqrt{3\sqrt{21}ac-10b^{3/2}}}{2\sqrt{42}a\sqrt[4]{b}\sinh\left(\frac{1}{7}\sinh^{-1}\left(\frac{\sqrt{3\sqrt{21}ac-10b^{3/2}}}{2\sqrt{2}b^{3/4}}\right)+\frac{i2\pi k}{7}\right)}}\ ,k=0,1,\dots,6. \tag{281d}$$

Certainly, we can find the coefficient for the x^6 term in equation (281c) by substituting x with x + b/(7a). However, the coefficients for the x^3, x^2, x and constant terms becomes progressively more complex [see Appendix 2-A-14]. Formula (281d) can be expressed in the sinht function in order to provide a complete solution.

E. First Method: Solution to Special Form the Eighth Degree Equation

In this section, we expand upon the solution obtained for the quartic equation mentioned either in (84b) or (85b) and utilize it to deduce the solution for the octic equation presented in the form,

$$a u^8 + b u^7 + c u^6 + d u^5 + e u^4 + f u^3 + \frac{c f^2}{d^2} u^2 + \frac{b f^3}{d^3} u + \frac{a f^4}{d^4} = 0 \quad (a \neq 0, d \neq 0) \tag{294}$$

Assuming f be a real number. By substituting x with u + f/u in (84a) and (84b), where u represents the roots of the octic equation, which gives

$$\begin{cases} a u^8 + b u^7 + (c + 4 a f) u^6 + (d + 3 b f) u^5 + (e + 6 a f^2 + 2 c f) u^4 + (3 b f^2 + d) u^3 \\ \qquad + (4 a f^3 + c f^2) u^2 + b f^2 u + a f^4 = 0 \quad (a \neq 0) \tag{295a} \\ u + \dfrac{f}{u} = x_{j+1} \Rightarrow u = \dfrac{1}{2} \left(x_{j+1} \pm \sqrt{x_{j+1}^2 - 4 f} \right), \tag{295b} \end{cases} \tag{295}$$

where x_{j+1} and the remaining dependent expressions in equation (84) remain unchanged.

We proceed the same procedure as before to construct the coefficients c, d, e, and f. However, due to the straightforward derivation of the formula, we won't delve into its detailed manipulation. Instead, we will provide a concise outline of how to derive the formula. By substituting c with c-4af, d with d-3bf, and e with e +2af²-2cf into equations (295) and (84), we arrive at the following derived equation in the special form:

$$a u^8 + b u^7 + c u^6 + d u^5 + e u^4 + f u^3 + \frac{c f^2}{d^2} u^2 + \frac{b f^3}{d^3} u + \frac{a f^4}{d^4} = 0 \quad (a \neq 0, d \neq 0)$$

And the solution can be expressed as

$$u_{1,\ldots,8} = \frac{1}{2} \left(x_{k+1} \pm \sqrt{x_{k+1}^2 - \frac{4 f}{d}} \right), \quad k = 0, 1, 2, 3, \tag{296}$$

where

$$x_{k+1} = -\rho - \frac{(b - 4 a \rho) T_k}{4 a} \left(1 - \cos\left[\frac{1}{2} \cos^{-1}\left(1 - \frac{32 a^2 (d^2 + 3 b d \rho^2 + 8 a f p - 4 a d \rho^3 - 3 b f - 2 c d \rho)}{d (b - 4 a \rho)^3 T_k^2} \right) \right] \right), \quad k = 0, 1. \tag{297}$$

$$x_{k+3} = -\rho - \frac{(b - 4 a \rho) T_k}{4 a} \left(1 + \cos\left[\frac{1}{2} \cos^{-1}\left(1 - \frac{32 a^2 (d^2 + 3 b d \rho^2 + 8 a f p - 4 a d \rho^3 - 3 b f - 2 c d \rho)}{d (b - 4 a \rho)^3 T_k^2} \right) \right] \right), \quad k = 0, 1. \tag{298}$$

157

$$T_k = 1 - \cos\left[\frac{1}{2}\cos^{-1}\left(\frac{2\left(b^3 d + 8a^2 d^2 - 8a^2 b f - 4abcd\right)}{d\left(b - 4a\rho\right)^3} - 1\right) + \pi k\right], \quad k = 0, 1. \tag{299}$$

$$\rho = \frac{D_1}{3dN} + \frac{2\sqrt{D_2^2 - 3d\,D_3\,D_4}}{3dN} * \cos\left(\frac{1}{3}\cos^{-1}\left(\frac{9dN\,D_2\,D_4 - 2D_2^3 - 27d^2 N^2 D_5}{2\left(D_2^2 - 3d\,D_4\,N\right)^{3/2}}\right)\right) \tag{300}$$

$$D_1 = 2abd^3 + 16a^2 d^2 e + cb^2 d^2 - 32a^3 f^2 - 10ab^2 df - 4ac^2 d^2$$

$$D_2 = 32a^3 f^2 + 10ab^2 df + 4ac^2 d^2 - 2abd^3 - 16a^2 d^2 e - cb^2 d^2$$

$$D_3 = b^3 d + 8a^2 d^2 - 8a^2 b f - 4abcd$$

$$D_4 = b^2 d^3 + 16a^2 d^2 f + 8abd^2 e - 3b^3 df - 4acd^3 - 32a^2 bf^2 - 4abcdf$$

$$D_5 = ad^4 + 7ab^2 f^2 + 2b^2 cdf - 6abd^2 f - b^2 d^2 e$$

$$N = b^3 d + 8a^2 d^2 - 8a^2 b f - 4abcd$$

Recall that it is important to note that the solution of equation (294) is derived in the composition form of cosine and its inverse. There is an analogous form involving hyperbolic cosine and its inverse, and there is no necessity to derive it separately. Both forms-utilizing cosine and hyperbolic cosine, respectively-offer a comprehensive solution to the equation.

Formula (296) generates a pair of conjugate roots for each value of k, resulting in a total of four pairs. Consequently, the formula provides eight roots, which can be explicitly expressed in terms of x_{k+1}, T_k and ρ:

$$u_{1,2} = \frac{1}{2}\left(x_1 \pm \sqrt{x_1^2 - \frac{4f}{d}}\right) \tag{301}$$

$$u_{3,4} = \frac{1}{2}\left(x_2 \pm \sqrt{x_2^2 - \frac{4f}{d}}\right) \tag{302}$$

$$u_{5,6} = \frac{1}{2}\left(x_3 \pm \sqrt{x_3^2 - \frac{4f}{d}}\right) \tag{303}$$

$$u_{7,8} = \frac{1}{2}\left(x_4 \pm \sqrt{x_4^2 - \frac{4f}{d}}\right) \tag{304}$$

where

$$x_1 = -\rho - \frac{(b - 4a\rho)T_0}{4a}\left(1 - \cos\left[\frac{1}{2}\cos^{-1}\left(1 - \frac{32a^2\left(d^2 + 3bd\rho^2 + 8afp - 4ad\rho^3 - 3bf - 2cd\rho\right)}{d\left(b - 4a\rho\right)^3 T_0^2}\right)\right]\right) \tag{296a}$$

158

$$x_2 = -\rho - \frac{(b-4a\rho)T_1}{4a}\left(1 - \cos\left[\frac{1}{2}\cos^{-1}\left(1 - \frac{32a^2(d^2+3bd\rho^2+8afp-4ad\rho^3-3bf-2cd\rho)}{d(b-4a\rho)^3T_1^2}\right)\right]\right) \qquad (296b)$$

$$x_3 = -\rho - \frac{(b-4a\rho)T_0}{4a}\left(1 + \cos\left[\frac{1}{2}\cos^{-1}\left(1 - \frac{32a^2(d^2+3bd\rho^2+8afp-4ad\rho^3-3bf-2cd\rho)}{d(b-4a\rho)^3T_0^2}\right)\right]\right) \qquad (296c)$$

$$x_4 = -\rho - \frac{(b-4a\rho)T_1}{4a}\left(1 + \cos\left[\frac{1}{2}\cos^{-1}\left(1 - \frac{32a^2(d^2+3bd\rho^2+8afp-4ad\rho^3-3bf-2cd\rho)}{d(b-4a\rho)^3T_1^2}\right)\right]\right) \qquad (296d)$$

$$T_0 = 1 - \cos\left[\frac{1}{2}\cos^{-1}\left(\frac{2(b^3d+8a^2d^2-8a^2bf-4abcd)}{d(b-4a\rho)^3}-1\right)\right] \qquad (299a)$$

$$T_1 = 1 + \cos\left[\frac{1}{2}\cos^{-1}\left(\frac{2(b^3d+8a^2d^2-8a^2bf-4abcd)}{d(b-4a\rho)^3}-1\right)\right] \qquad (299b)$$

F. Second Method: Solution to Special Form of the Eighth Degree Equation

We present an alternative approach to solving the quartic equation in the given form,

$$x^8 + ax^7 + bx^6 + cx^5 + dx^4 + ex^3 + \frac{be^2}{c^2}x^2 + \frac{ae^3}{c^3}x + \frac{e^4}{c^4} = 0, \quad (c \neq 0). \qquad (300)$$

Our second method involves initially solving the quartic equation in its factored form. Once we have obtained the solution, we proceed to apply the same procedure to the First Method.

Assuming that a, b, c, m, n, p and q be real numbers such that the following quartic equation can be factorised to give

$$x^4 + ax^3 + bx^2 + cx + d = (x^2 + mx + n)(x^2 + px + r)$$

$$= x^4 + (m+p)x^3 + (n+r+mp)x^2 + (mr+np)x + nr = 0 \qquad (301)$$

Each side of (301) contains a quartic equation, and our objective is to determine the values of m, n, p and r in terms of a, b, c and d. To simplify the solution process, we convert both sides of equation (301) to the depressed equation by setting the coefficients of the x^3 term to 0. This is achieved by equating m+p = a = 0, which implies that m = -p and gives:

$$x^4 + (n+r-p^2)x^2 + (n-r)px + np = 0 \qquad (302)$$

The solution to the quadratic equation, $x^2 + mx + n = 0$, is

$$x_{k+1} = -\frac{m}{2}\left(1 - \cos\left[\frac{1}{2}\cos^{-1}\left(\frac{m^2 - 8n}{m^2}\right) + k\pi\right]\right), k = 0, 1.$$

$$= \frac{p}{2}\left(1 - \cos\left[\frac{1}{2}\cos^{-1}\left(\frac{p^2 - 8n}{p^2}\right) + k\pi\right]\right), k = 0, 1. \tag{303}$$

And the solution to another quadratic equation, $x^2 + px + r = 0$, is

$$x_{k+3} = -\frac{p}{2}\left(1 - \cos\left[\frac{1}{2}\cos^{-1}\left(\frac{p^2 - 8r}{p^2}\right) + k\pi\right]\right), k = 0, 1. \tag{304}$$

We determine n, p and r as follows:

1. By equating the coefficients between the equation on the left-hand side of (301) and equation (302), we obtain:

$$\begin{cases} m + p = a = 0 & \Rightarrow m = -p & (305) \\ n + r + mp = b & \Rightarrow n + r = b + p^2 & (306) \\ mr + np = c & \Rightarrow n - r = \dfrac{c}{p} & (307) \\ nr = d & & (309) \end{cases}$$

From (306) and (307), solving for n and r gives

$$n = \frac{1}{2}\left(b + p^2 + \frac{c}{p}\right)$$

and

$$r = \frac{1}{2}\left(b + p^2 - \frac{c}{p}\right)$$

2. By substituting the values of n and r into (309), we have

$$\frac{1}{2}\left(b + p^2 + \frac{c}{p}\right) \cdot \frac{1}{2}\left(b + p^2 - \frac{c}{p}\right) = d$$

After multiplying through and simplifying, we obtain the sextic equation in p which is similar to Ramanujan's method [7] to solve for the quartic equation.

160

$$p^6 + 2bp^4 + (b^2 - 4d)p^2 - c^2 = 0.$$ (310)

3. Applying the cubic formula to (310) gives the cubic roots, which are given by

$$p^2 = -\frac{2b}{3} + \frac{2}{3}\sqrt{b^2 + 12d}\cos\left(\frac{1}{3}\cos^{-1}\frac{27c^2 + 2b^3 - 72bd}{2(b^2 + 12d)^{3/2}}\right).$$ (311)

4. By substituting a = 0 in equation (300), n into formula (303) and r into formula (304), we obtain the roots to the depressed quartic equation as follows:

$$x^4 + bx^2 + cx + d = 0$$ (312)

$$x_{k+1} = \frac{p}{2}\left(1 - \cos\left[\frac{1}{2}\cos^{-1}\left(-\frac{4c}{p^3} - \frac{4b}{p^2} - 3\right) + \frac{k\pi}{2}\right]\right)$$

$$= \frac{p}{2}\left(1 + \cos\left[\frac{1}{2}\cos^{-1}\left(\frac{4c}{p^3} + \frac{4b}{p^2} + 3\right) + \frac{k\pi}{2}\right]\right), k = 0, 1,$$ (313)

and

$$x_{k+3} = -\frac{p}{2}\left(1 - \cos\left[\frac{1}{2}\cos^{-1}\left(\frac{4c}{p^3} - \frac{4b}{p^2} - 3\right) + \frac{k\pi}{2}\right]\right), k = 0, 1,$$ (314)

where p is given by (311).

The depressed equation (312) can be transformed into standard form by replacing x with x + b/(4a) and then following the same steps as illustrated previously to determine other coefficients, which results in.

$$x^4 + ax^3 + bx^2 + cx + d = 0$$ (315)

$$x_{k+1} = -\frac{a}{4} + \frac{p}{2}\left(1 - \cos\left[\frac{1}{2}\cos^{-1}\left(\frac{4ab - 8c - a^3}{2p^3} - \frac{8b - 3a^2}{2p^2} - 3\right) + \frac{k\pi}{2}\right]\right), k = 0, 1.$$ (316)

$$x_{k+3} = -\frac{a}{4} - \frac{p}{2}\left(1 + \cos\left[\frac{1}{2}\cos^{-1}\left(\frac{4ab - 8c - a^3}{2p^3} + \frac{8b - 3a^2}{2p^2} + 3\right) + \frac{k\pi}{2}\right]\right), k = 0, 1.$$ (317)

where p is given by

$$p^2=\frac{a^2}{4}-\frac{2b}{3}+\frac{2}{3}\sqrt{b^2+12d-3ac}\cdot\cos\left(\frac{1}{3}\cos^{-1}\frac{2b^3+27\left(c^2+a^2d\right)-9abc-72bd}{2\left(b^2+12d-3ac\right)^{3/2}}+\frac{2\pi k}{3}\right) \qquad (318)$$

Since the leading term in equation (315) has a coefficient of 1, the octic formula will exhibit slight variations compared to the solution from the first method. After substituting x with (x/e + 1/x) into equation (315), and following the same steps as indicated in the first method to determine the coefficients of the octic equation, we derive the following results:

$$x^8+ax^7+bx^6+cx^5+dx^4+ex^3+\frac{be^2}{c^2}x^2+\frac{ae^3}{c^3}x+\frac{e^4}{c^4}=0 \quad (c\neq0) \qquad (319)$$

$$x_{1,\ldots,8}=\frac{\beta_{k+1}e}{2c}\pm\frac{1}{2c}\sqrt{\beta_{k+1}^2e^2-4ce} \qquad (320)$$

$$\beta_{k+1}=-\frac{ac}{4e}+\frac{p}{2}\left(1-\cos\left[\frac{1}{2}\cos^{-1}\left(\frac{4abc^3+8ac^2e-8bc^2ep+32ce^2p-8c^4+3a^2c^2ep-a^3c^3-6e^3p^3}{2e^3p^3}\right)+k\pi\right]\right),k=0,1. \qquad (321)$$

$$p^2=\frac{3a^2c^2+32ec-8bc^2}{12e^2}+\frac{2c}{3e^2}\sqrt{9a^2ce+b^2c^2+12c^2d+40e^2-3ac^3-32bce*}$$

$$\cos\left(\frac{1}{3}\cos^{-1}\frac{189a^2ce^2+27a^2c^3d+120b^2c^2e+2b^3c^3+288c^2de+27c^5+448e^3-9abc^4-126ac^3e-624bce^2-72bc^3d-27a^2bc^2e}{2\left(9a^2ce+b^2c^2+12c^2d+40e^2-3ac^3-32bce\right)^{3/2}}\right) \qquad (322)$$

Formula (320) provides the roots to octic equation (319). It is dependent on the values of β_1 and β_2 from (321), as well as the real root of the cubic equation defined by (322), denoted by p. Once p is determined, there are always two real values of p: $+p$ and $-p$, due to the square root. This is a factor that contributes to formula (320) having eight roots. Therefore, the roots to equation (319) can be explicitly expressed using formula (320), as shown below:

Define the values for D, T, T₂:

$$D=9a^2ce+b^2c^2+12c^2d+40e^2-3ac^3-32bce \qquad (323)$$

$$T=\frac{189a^2ce^2+27a^2c^3d+120b^2c^2e+2b^3c^3+288c^2de+27c^5+448e^3-9abc^4-126ac^3e-624bce^2-72bc^3d-27a^2bc^2e}{2D^{3/2}}$$

$$T_2=\frac{4abc^3+8ac^2e-8bc^2ep+32ce^2p-8c^4+3a^2c^2ep-a^3c^3-6e^3p^3}{2e^3p^3} \qquad (325)$$

Determine p, β_1 and β_2:

$$\rho = \begin{cases} \pm\sqrt{\dfrac{3a^2c^2+32ec-8bc^2}{12e^2}+\dfrac{2c}{3e^2}\sqrt{D}*\cos\left(\dfrac{1}{3}\cos^{-1}(T)\right)},\; -1\le T\le 1 & (326\,a) \\[3ex] \pm\sqrt{\dfrac{3a^2c^2+32ec-8bc^2}{12e^2}+\dfrac{2c}{3e^2}\sqrt{D}*\cosh\left(\dfrac{1}{3}\cosh^{-1}(T)\right)},\; T\ge 1 & (326\,b) \\[3ex] \pm\sqrt{\dfrac{3a^2c^2+32ec-8bc^2}{12e^2}-\dfrac{2c}{3e^2}\sqrt{D}*\cosh\left(\dfrac{1}{3}\cosh^{-1}(-T)\right)},\; T\le -1 & (326\,c) \end{cases} \qquad (326)$$

There are always two values, $\pm\rho$, that exist since ρ is a real root of cubic equation (322).

$$\beta_1 = \begin{cases} -\dfrac{ac}{4e}+\dfrac{\rho}{2}\left(1-\cos\left(\dfrac{1}{2}\cos^{-1}(T_2)\right)\right),\; -1\le T_2\le 1 \\[3ex] -\dfrac{ac}{4e}+\dfrac{\rho}{2}\left(1-\cosh\left(\dfrac{1}{2}\cosh^{-1}(T_2)\right)\right),\; T_2\ge 1 \\[3ex] -\dfrac{ac}{4e}+\dfrac{\rho}{2}\left(1+\sinh\left(\dfrac{1}{2}\cosh^{-1}(-T_2)\right)\right),\; T_2\le -1 \end{cases}$$

$$\beta_2 = \begin{cases} -\dfrac{ac}{4e}+\dfrac{\rho}{2}\left(1+\cos\left(\dfrac{1}{2}\cos^{-1}(T_2)\right)\right),\; -1\le T_2\le 1 \\[3ex] -\dfrac{ac}{4e}+\dfrac{\rho}{2}\left(1+\cosh\left(\dfrac{1}{2}\cosh^{-1}(T_2)\right)\right),\; T_2\ge 1 \\[3ex] -\dfrac{ac}{4e}+\dfrac{\rho}{2}\left(1-\sinh\left(\dfrac{1}{2}\cosh^{-1}(-T_2)\right)\right),\; T_2\le -1 \end{cases}$$

Then the roots of equation (319) are

$$x_{1,2}=\frac{\beta_1 e}{2c}\pm\frac{1}{2c}\sqrt{\beta_1^2 e^2-4ce}$$

$$x_{3,4}=\frac{\beta_2 e}{2c}\pm\frac{1}{2c}\sqrt{\beta_2^2 e^2-4ce},$$

note, the positive value of ρ in (326) is used.

and

$$x_{5,6}=\frac{\beta_1 e}{2c}\pm\frac{1}{2c}\sqrt{\beta_1^2 e^2-4ce}$$

$$x_{7,8}=\frac{\beta_2 e}{2c}\pm\frac{1}{2c}\sqrt{\beta_2^2 e^2-4ce},$$

the negative value of ρ in (326) is used.

G. Suggesting New Elementary Functions

During the exploration of solving quadratic, cubic and quartic equations by trigonometric approach, we propose a side suggestion, which is implemented in this book only. We suggest introducing potential new elementary functions in mathematics, perhaps named **cosht** (or an alternative), **sinht** and **tanht**. The cosht function represents a transformative link between the combination of composing cosine and its inverse and composing hyperbolic cosine and its inverse. The name **cosht** derives from **cos** for cosine and **cosh** for hyperbolic cosine, symbolizing its transformative role within these functions. Similarly, the name **sinht** derives from **sin** for sine and **sinh** for hyperbolic sine, and **tanht** derives from **tan** for tangent and **tanh** for hyperbolic tangent. The letter **t** in cosht, sinht or tanht signifies its purpose as a transformation between the composing trigonometric functions and composing hyperbolic trigonometric functions during the solving process due to the constraint of their inverse function's domains. It would essentially function similarly but with distinct domains.

The functions *cosht, sinht* and *tanht* can be defined as follows:

$$\mathrm{cosht}_{n,k}(x)=\begin{cases}\cos\left(\dfrac{1}{n}\cos^{-1}(x)+\dfrac{2\pi k}{n}\right) & -1\le x\le 1 \quad (350a)\\[2mm]\cosh\left(\dfrac{1}{n}\cosh^{-1}(x)+\dfrac{i2\pi k}{n}\right) & x\ge 1 \quad (350b)\\[2mm]\cosh\left(\dfrac{1}{n}\cosh^{-1}(-x)-\dfrac{i\pi(2k+1)}{n}\right) & x\le -1 \quad (350c)\\[2mm]\cos\left(\dfrac{1}{n}\cos^{-1}(ix)+\dfrac{2\pi k}{n}\right)=i\sinh\left(\dfrac{1}{n}\sinh^{-1}(x)+\dfrac{i\pi(4k-n+1)}{2n}\right) & -\infty<x<\infty \quad (350d)\\[2mm]\cosh\left(\dfrac{1}{n}\cosh^{-1}(ix)+\dfrac{i2\pi k}{n}\right)=i\sinh\left(\dfrac{1}{n}\sinh^{-1}(x)+\dfrac{i\pi(4k-n+1)}{2n}\right) & -\infty<x<\infty \quad (350e)\\[2mm]\text{for } k=0,1,\ldots,n-1.\end{cases} \quad (350)$$

The cosht function has domain (-∞, ∞).

$$\text{sinht}_{n,k}(x)=\begin{cases} \sin\left(\dfrac{1}{n}\sin^{-1}(x)+\dfrac{2\pi k}{n}\right)=\begin{cases} \sin\left(\dfrac{1}{n}\sin^{-1}(x)+\dfrac{2\pi k}{n}\right) & -1\le x\le 1 \quad (351a) \\[2mm] \cosh\left(\dfrac{1}{n}\cosh^{-1}(x)+\dfrac{i\pi(4k-n+1)}{2n}\right) & x\ge 1 \quad (351b) \\[2mm] \cosh\left(\dfrac{1}{n}\cosh^{-1}(-x)-\dfrac{i\pi(4k-n-1)}{2n}\right) & x\le -1 \quad (351c) \end{cases} \\[8mm] \sin\left(\dfrac{1}{n}\sin^{-1}(ix)+\dfrac{2\pi k}{n}\right)=\left\{ i\sinh\left(\dfrac{1}{n}\sinh^{-1}(x)+\dfrac{2i\pi k}{n}\right)\right. & -\infty<x<\infty \quad (351d) \\[4mm] \sinh\left(\dfrac{1}{n}\sinh^{-1}(x)+\dfrac{i2\pi k}{n}\right) & -\infty<x<\infty \quad (351e) \\[6mm] \sinh\left(\dfrac{1}{n}\sinh^{-1}(ix)+\dfrac{i2\pi k}{n}\right)=\begin{cases} i\sin\left(\dfrac{1}{n}\sin^{-1}(x)+\dfrac{2\pi k}{n}\right) & -1\le x\le 1 \quad (351f) \\[2mm] i\cosh\left(\dfrac{1}{n}\cosh^{-1}(x)+\dfrac{i\pi(4k-n+1)}{2n}\right) & x\ge 1 \quad (351g) \\[2mm] i\cosh\left(\dfrac{1}{n}\cosh^{-1}(-x)-\dfrac{i\pi(4k-n-1)}{2n}\right) & x\ge 1 \quad (351h) \end{cases} \\[8mm] \text{for } k=0,1,\dots,n-1. \end{cases} \quad (351)$$

The sinht function has domain $(-\infty, \infty)$.

$$\text{tanht}_{n,k}(x)=\begin{cases} \tan\left(\dfrac{1}{n}\tan^{-1}(x)+\dfrac{\pi k}{n}\right)=\left\{\tan\left(\dfrac{1}{n}\tan^{-1}(x)+\dfrac{\pi k}{n}\right)\right. & -\infty<x<\infty \quad (352a) \\[6mm] \tan\left(\dfrac{1}{n}\tan^{-1}(ix)+\dfrac{\pi k}{n}\right)=\begin{cases} i\tanh\left(\dfrac{1}{n}\tanh^{-1}(x)+\dfrac{i\pi k}{n}\right) & -1<x<1 \quad (352b) \\[2mm] i\tanh\left(\dfrac{1}{n}\coth^{-1}(x)+\dfrac{i\pi(2k+1)}{n}\right) & x<-1\,\text{or}\,x>1 \quad (352c) \end{cases} \\[8mm] \tanh\left(\dfrac{1}{n}\tanh^{-1}(x)+\dfrac{i\pi k}{n}\right) & -1<x<1 \quad (352d) \\[4mm] \tanh\left(\dfrac{1}{n}\tanh^{-1}(ix)+\dfrac{i\pi k}{n}\right)=i\tan\left(\dfrac{1}{n}\tan^{-1}(x)+\dfrac{\pi k}{n}\right) & -\infty<x<\infty \quad (352e) \\[4mm] k=0,1,\dots,n-1, \end{cases} \quad (352)$$

The tanht function has domain $(-\infty\ -1) \cup (-1, 1) \cup (1, \infty)$.

where

- n be a non-negative integer ($n \in \mathbb{N}$) representing the highest power for the equation.

- k be an integer ($k \in \mathbb{Z}$) indicating the number of roots for the equation.

Specifically, if k = 0, there is one root; if k = 1, there are two roots; if k = 2, there are three roots, and so on.

The domains of cosht, sinht and tanht are all real numbers. However, the domain of tanht is all real numbers except for x = 1 or x = -1. If the input of these functions is in the form of ix, then the product of the function and a radical expression may have at least one real number. If the input value of these functions takes the form of a + iβx, where a and β are real, then either $\cos((1/n)\cos^{-1}(x))$ or $\cosh((1/n)\cosh^{-1}(x))$, for example, is sufficient for cosht function to compute the roots of the equation, according to the modern calculator. In that case, the general solution formula expressed in complex number becomes very complicated. Although our derived solution formulas for quartic, quintic, and septic equations in prior sections hold for almost all cases in computing, we do not consider deriving the solution formulas when the coefficients of the equation are of complex form in this book. Note that the relations, for example, $\sin(ix) = i\sinh(x)$, $\cos(ix)=\cosh(x)$ and $\tan(ix)=i\tanh(x)$, are not restricted to the principal branch. They extend across the entire complex plane.

Next, we provide the proofs for (350), (351) and (352), and present example.

Proofs

1. To prove formula (350b), we can follow the same approach as [2-A-6] and apply it for general n instead.

2. Formula (350c) follows by replacing x with -x in (350b) and applying the equality $\cosh^{-1}(-x) = i\pi - \cosh^{-1}(x)$.

3. To prove (350d), we use this identity,

$$\cos^{-1}(x)=\frac{\pi}{2}-\sin^{-1}(x).$$

Replacing x by ix gives

$$\cos^{-1}(ix)=\frac{\pi}{2}-i\sinh^{-1}(x)$$

By substitution, we have

$$\cos\left(\frac{1}{n}\cos^{-1}(ix)+\frac{2\pi k}{n}\right)=\cos\left(\frac{\pi}{2n}-\frac{i}{n}\sinh^{-1}(x)+\frac{2\pi k}{n}\right) \quad -\infty<x<\infty$$

$$=\cos\left(-\frac{i}{n}\sinh^{-1}(x)+\frac{\pi(4k+1)}{2n}\right) \quad -\infty<x<\infty$$

$$= \cos\left(\frac{\pi}{2} - \frac{i}{n}\sinh^{-1}(x) + \frac{\pi(4k+1)}{2n} - \frac{\pi}{2}\right) \quad -\infty < x < \infty$$

$$= \sin\left(\frac{i}{n}\sinh^{-1}(x) - \frac{\pi(4k-n+1)}{2n}\right) \quad -\infty < x < \infty$$

$$= i\sinh\left(\frac{1}{n}\sinh^{-1}(x) + \frac{i\pi(4k-n+1)}{2n}\right) \quad -\infty < x < \infty$$

4. To prove formula (351b), we use the definition of the inverse sine defined in the main branch of the complex plane that is expressed as follows:

$$\sin^{-1}x = i\ln\left(\sqrt{1-x^2} - ix\right), -1 \leq x \leq 1.$$

If x ≥ 1, we have

$$\sin^{-1}x = i\ln\left(i\sqrt{x^2-1} - ix\right)$$

$$= i\ln(i) + i\ln\left(\sqrt{x^2-1} - x\right)$$

By applying the equality, $\left(\sqrt{x^2-1} - x\right)\left(\sqrt{x^2-1} + x\right) = -1$, the inverse sine of x becomes

$$\sin^{-1}x = i\ln(i) + i\ln\left(\frac{-1}{\sqrt{x^2-1} + x}\right)$$

$$= i\ln(i) + i\ln(-1) - i\ln\left(\sqrt{x^2-1} + x\right)$$

We have $\ln(i) = i\pi/2, \ln(-1) = i\pi$ and $\ln(\sqrt{x^2-1} + x) = \cosh^{-1}(x)$. Hence, by substitution these in the expression above, we get

$$\sin^{-1}x = -\frac{3\pi}{2} - i\cosh^{-1}(x).$$

Next, we consider the expression,

$$\sin\left(\frac{1}{n}\sin^{-1}(x) + \frac{2\pi k}{n}\right) = \sin\left(\frac{1}{n}\left(-\frac{3\pi}{2} - i\cosh^{-1}(x)\right) + \frac{2\pi k}{n}\right)$$

$$= \sin\left(-\frac{i}{n}\cosh^{-1}(x) + \frac{4\pi k - 3\pi}{2n}\right)$$

167

$$= \sin\left(\frac{\pi}{2} - \frac{i}{n}\cosh^{-1}(x) + \frac{4\pi k - 3\pi}{2n} - \frac{\pi}{2}\right)$$

$$= \cos\left(\frac{i}{n}\cosh^{-1}(x) + \frac{4\pi k - \pi n - 3\pi}{2n}\right).$$

By replacing k by k+ 1 and factoring out, which gives

$$= \cosh\left(\frac{1}{n}\cosh^{-1}(x) + \frac{i\pi(4k - n + 1)}{2n}\right)$$

5. To prove (351c), we follow proof #2. Additionally, if k is replaced by -k in (351c), the expression on the right-hand side is rewritten a little differently as

$$\sinh t_{n,k}(x) = \cosh\left(\frac{1}{n}\cosh^{-1}(-x) + \frac{i\pi(4k + n + 1)}{2n}\right), \quad x \le -1, k = 0, 1, \ldots, n-1.$$

6. Replace x by ix and use $\sin^{-1}(ix) = i\sinh^{-1}(x)$. Formula (351d) follows after replacing k by -k and a little simplification.

7. To prove (351g) and (351h), we follow the same proofs #1 and #2.

8. Formula (352b) follows after replacing x by ix and k with -k.

9. For x < -1 or x > 1, result (352c) is expected by applying the equality

$$\tanh^{-1}x = \frac{i\pi}{2} + \coth^{-1}x.$$

10. To prove (352e), we use two identities $\tanh^{-1}(ix) = i\tan^{-1}(x)$ and $\tanh(ix) = i\tan(x)$, the proof follows after a little simplification.

We note that formula (352e) can also be expressed in terms of $\coth((1/n)\coth^{-1}(x))$ as follows:

$$i\tanh\left(\frac{1}{n}\coth^{-1}(x) + \frac{i\pi(2k+1)}{n}\right) = i\tanh\left(\frac{i\pi}{2} + \frac{1}{n}\coth^{-1}(x) + \frac{i\pi(2k+1)}{n} - \frac{i\pi}{2}\right)$$

$$= i\coth\left(\frac{1}{n}\coth^{-1}(x) + \frac{i\pi(4k - n + 2)}{n}\right).$$

Furthermore, if n is an odd number, and we express n as 2m+1, with m∈ℕ, the phase shift for the hyperbolic cosine is identical for x ≥ 1 or x ≤ -1,

$$\cosh t_{2m+1,k}(x)=\begin{cases}+\cosh\left(\dfrac{1}{2m+1}\cosh^{-1}(x)+\dfrac{i\,2\pi k}{2m+1}\right) & \text{if } x\geq1 \\[2ex] -\cosh\left(\dfrac{1}{2m+1}\cosh^{-1}(-x)+\dfrac{i\,2\pi k}{2m+1}\right) & \text{if } x\leq-1. \\[2ex] m=0,1,2,\dots \\ k=0,1,2,\dots,2m. \end{cases}\qquad(353)$$

Conclusion

The new functions sinht, cosht and tanht acts like a transformative link or a gateway that establishes a connection among these types of functions via a single atomic function, helps simplifying the process of choosing a formula and computes the roots of an equation with ease. The integration of the trigonometric and hyperbolic trigonometric functions offers a unified element that fulfills the necessary roles of the new function operating on both real and complex numbers without having constraint by domains or any limitations except some singular points. This can be helpful in solving problems that involve both real and imaginary numbers, especially in higher order polynomials. It can also enhance computational convenience in the future. From now on, when a solution formula is expressed in either function (350), (351) or (352), it implies that the formula provides a complete solution.

In addition, it has come to our attention that definition (350) bears resemblance to the Chebyshev polynomial of the first kind. Nevertheless, it deviates as a result of a unique value of (n), which we define as its reciprocal, (1/n), and the phase angles present in the cosine or hyperbolic cosine functions.

Example. Rewrite solution formula (281b) for the septic equation using the sinht function and solution formula (320) for the octic equation using the cosht function.

Solution

1. The solution formula for the special septic equation can be expressed in terms of sinht function as

$$\begin{cases}a x^{7}+b x^{6}+c x^{4}+\dfrac{c^{2}}{4b}x^{2}+\dfrac{c^{3}}{56b^{2}}=0 \quad (a\neq0,b\neq0) \\[3ex] x_{k+1}=-\dfrac{\sqrt{c}}{2\sqrt{2b}\,\sinh t_{7,k}\left(\dfrac{7a\sqrt{c}}{2b\sqrt{2b}}\right)} \quad,\ k=0,1,\dots,6.\end{cases}\qquad(354)$$

2. The solution formula for the special octic equation, expressed in terms of cosht

function, is

$$x^8+ax^7+bx^6+cx^5+dx^4+ex^3+\frac{be^2}{c^2}x^2+\frac{ae^3}{c^3}x+\frac{e^4}{c^4}=0 \quad (c\neq0) \tag{355a}$$

$$x_{1,\ldots,8}=\frac{\beta_{k+1}e}{2c}\pm\frac{1}{2c}\sqrt{\beta_{k+1}^2e^2-4ce} \tag{355b}$$

$$\beta_{k+1}=-\frac{ac}{4e}+\frac{p}{2}\left(1-\cosh t_{2,k}\left(\frac{4abc^3+8ac^2e-8bc^2ep+32ce^2p-8c^4+3a^2c^2ep-a^3c^3-6e^3p^3}{2e^3p^3}\right)\right),k=0,1 \tag{355c}$$

$$p^2=\frac{3a^2c^2+32ec-8bc^2}{12e^2}+\frac{2c}{3e^2}\sqrt{9a^2ce+b^2c^2+12c^2d+40e^2-3ac^3-32bce*}$$

$$\cosh t_{3,k}\left(\frac{189a^2ce+27a^2c^3d+120b^2c^2e+2b^3c^3+288c^2de+27c^5+448e^3-9abc^4-126ac^3e-624bce^2-72bc^3d-27a^2bc^2e}{2\left(9a^2ce+b^2c^2+12c^2d+40e^2-3ac^3-32bce\right)^{3/2}}\right),$$

$$k=0,1,2. \tag{355d}$$

$$(355)$$

Then the roots of equation (355a) are

$$x_{1,2}=\frac{\beta_1 e}{2c}\pm\frac{1}{2c}\sqrt{\beta_1^2 e^2-4ce} \tag{355e}$$

$$x_{3,4}=\frac{\beta_2 e}{2c}\pm\frac{1}{2c}\sqrt{\beta_2^2 e^2-4ce}, \tag{355f}$$

where the positive value of ρ is used.

and

$$x_{5,6}=\frac{\beta_1 e}{2c}\pm\frac{1}{2c}\sqrt{\beta_1^2 e^2-4ce} \tag{355g}$$

$$x_{7,8}=\frac{\beta_2 e}{2c}\pm\frac{1}{2c}\sqrt{\beta_2^2 e^2-4ce}, \tag{355h}$$

where the negative value of ρ is used.

Using (355) with a = 0 and b = 0 results in *depressed-like* form of the octic equation,

$$\left|\begin{array}{l} x^8+c\,x^5+d\,x^4+e\,x^3+\dfrac{e^4}{c^4}=0 \quad (c\neq 0) \\[3mm] x_{1,\dots,8}=\dfrac{\beta_{k+1}e}{2c}\pm\dfrac{1}{2c}\sqrt{\beta_{k+1}^2e^2-4ce} \\[3mm] \beta_{k+1}=\dfrac{p}{2}\left(1-\cosh t_{2,k}\left(\dfrac{16\,c\,e^2\,p-4\,c^4-3\,e^3\,p^3}{e^3\,p^3}\right)\right),\ k=0,1 \\[3mm] p^2=\dfrac{8\,e\,c}{3\,e^2}+\dfrac{4\,c}{3\,e^2}\sqrt{3\,c^2d+10\,e^2}*\cosh t_{3,k}\left(\dfrac{288\,c^2d\,e+27\,c^5+448\,e^3}{16\left(3\,c^2d+10\,e^2\right)^{3/2}}\right),\ k=0,1,2. \end{array}\right.$$

(356 a)

(356 b)

(356 c)

(356 d)

(356)

The roots of equation (356a) can be explicitly expressed as (355e), (355f), (355h) and (355g). Note, we can express all solution formulas (355) and (356) to radical form by using (46b).

If we replace e by ec, then (356) can be rewritten a little differently as

$$\left|\begin{array}{l} x^8+c\,x^5+d\,x^4+ce\,x^3+e^4=0 \\[3mm] x_{1,\dots,8}=\dfrac{\beta_{k+1}e}{2}\pm\dfrac{1}{2}\sqrt{\beta_{k+1}^2e^2-4e} \\[3mm] \beta_{k+1}=\dfrac{p}{2}\left(1-\cosh t_{2,k}\left(\dfrac{16\,e^2\,p-4\,c-3\,e^3\,p^3}{e^3\,p^3}\right)\right),\ k=0,1 \\[3mm] p^2=\dfrac{8\,e}{3\,e^2}+\dfrac{4}{3\,e^2}\sqrt{3\,d+10\,e^2}*\cosh t_{3,k}\left(\dfrac{288\,d\,e+27\,c^2+448\,e^3}{16\left(3\,d+10\,e^2\right)^{3/2}}\right),\ k=0,1,2. \end{array}\right.$$

(357 a)

(357 b)

(357 c)

(357 d)

(357)

H. General Solution to Special Form of the (2n+1)th Degree Equations

We have already seen the forms of the quintic equation (228a) and septic equation (274a), both of which are solvable equations. In this section, we will provide a general solution formula for (2n+1)th degree equation. Let a and b are real numbers satisfying the condition ab = 1. We can observe the following results:

1. $a^3+b^3=(a+b)^3-3(a+b)$

(358)

By setting u = a + b and $a^3 + b^3 = m$, which gives a system of equations:

$$\left\{\begin{array}{l} u^3-3u-m=0 \\ a^3+b^3=m \\ (ab)^3=1 \end{array}\right.$$

(359 a)

(359 b)

(359 c)

(359)

Solve for a³ and b³ from (359b) and (359c), we obtain

$$a^3 = \frac{m}{2} + \frac{1}{2}\sqrt{m^2 - 4} \quad \text{and} \quad b^3 = \frac{m}{2} - \frac{1}{2}\sqrt{m^2 - 4}$$

Therefore, the root of equation (359a) is given by

$$u = a + b = \left(\frac{m}{2} + \frac{1}{2}\sqrt{m^2 - 4}\right)^{1/3} + \left(\frac{m}{2} - \frac{1}{2}\sqrt{m^2 - 4}\right)^{1/3}$$

$$= \begin{cases} 2\cosh\left(\frac{1}{3}\cosh^{-1}\left(\frac{m}{2}\right)\right) & m \geq 1 \\ 2\cos\left(\frac{1}{3}\cos^{-1}\left(\frac{m}{2}\right)\right) & -1 \leq m \leq 1 \end{cases} \tag{360}$$

If we replace b with -b, we can obtain u = a-b, which provides a solution to equation (358) in the form of either sine and its inverse -2sin((1/3)sin⁻¹(m/2)) or hyperbolic sine and its inverse -2sinh((1/3)sinh⁻¹(m/2)).

2. We now consider

$$(a+b)^5 = a^5 + b^5 + 5(a^3 + b^3) + 10(a+b)$$
$$= a^5 + b^5 + 5(a+b)^3 - 5(a+b). \quad \text{(Use (358))}$$

Hence, we get

$$a^5 + b^5 = (a+b)^5 - 5(a+b)^3 + 5(a+b) \tag{361}$$

Similarly to #1, setting u = a + b and a⁵ + b⁵ = m, we obtain the equation

$$u^5 - 5u^3 + 5u - m = 0 \tag{362}$$

and solve for a⁵ and b⁵

$$a^5 = \frac{m}{2} + \frac{1}{2}\sqrt{m^2 - 4} \quad \text{and} \quad b^5 = \frac{m}{2} - \frac{1}{2}\sqrt{m^2 - 4}.$$

The solution to equation (362) is

$$u = a + b = \left(\frac{m}{2} + \frac{1}{2}\sqrt{m^2 - 4}\right)^{1/5} + \left(\frac{m}{2} - \frac{1}{2}\sqrt{m^2 - 4}\right)^{1/5}$$

$$= \begin{cases} 2\cosh\left(\dfrac{1}{5}\cosh^{-1}\left(\dfrac{m}{2}\right)\right) & m \geq 1 \\[3mm] 2\cos\left(\dfrac{1}{5}\cos^{-1}\left(\dfrac{m}{2}\right)\right) & -1 \leq m \leq 1. \end{cases} \tag{363}$$

Note that the term $2\pi k$ or $i2\pi k$ is not included in the solution formula (363) during the solving process; it will be added when we derive the final solution.

3. Continuing the pattern above (#1 and #2) for the sum 7th powers ($a^7 + b^7$), we obtain the corresponding result,

$$(a+b)^7 = a^7 + b^7 + 7\left(a^5 + b^5\right) + 21\left(a^3 + b^3\right) + 35\left(a+b\right)$$

$$= a^7 + b^7 + 7\left((a+b)^5 - 5(a+b)^3 + 5(a+b)\right) + 21\left((a+b)^3 - 3(a+b)\right) + 35(a+b) \quad \text{(Use (358), (361))}$$

Hence, we deduce that

$$a^7 + b^7 = (a+b)^7 - 7(a+b)^5 + 14(a+b)^3 - 7(a+b). \tag{364}$$

Setting $u = a + b$ and $a^7 + b^7 = m$, we obtain

$$u^7 - 7u^5 + 14u^3 - 7u - m = 0 \tag{365}$$

and get the solution

$$u = a + b = \left(\frac{m}{2} + \frac{1}{2}\sqrt{m^2 - 4}\right)^{1/7} + \left(\frac{m}{2} - \frac{1}{2}\sqrt{m^2 - 4}\right)^{1/7}$$

$$= \begin{cases} 2\cosh\left(\dfrac{1}{7}\cosh^{-1}\left(\dfrac{m}{2}\right)\right) & m \geq 1 \\[3mm] 2\cos\left(\dfrac{1}{7}\cos^{-1}\left(\dfrac{m}{2}\right)\right) & -1 \leq m \leq 1. \end{cases} \tag{366}$$

Analyzing the coefficients generated by equations (359a), (362), and (365), we observe a similarity to the triangular coefficients of Chebyshev polynomials [12] when replacing u with u/2 in the equations and multiplying each equation by 2. To generalize the solution to this equation type, let a, b and p be arbitrary so that it satisfies the condition ab = p. Using the same procedure outlined in #1, #2, and #3, we derive formulas in the following patterns:

➢ Third degree equation and its solution

173

$$\begin{cases} u^3-3\,p\,u-m=0 \\ u=2\sqrt{p}\cosh\left(\dfrac{1}{3}\cosh^{-1}\dfrac{m}{2\,p^{3/2}}\right),m\geq 1 \\ u=2\sqrt{p}\cos\left(\dfrac{1}{3}\cos^{-1}\dfrac{m}{2\,p^{3/2}}\right),-1\leq m\leq 1 \end{cases} \tag{367}$$

> De Moivre's quintic equation and its solution

$$\begin{cases} u^5-5\,p\,u^3+5\,p^2u-m=0 \\ u=2\sqrt{p}\cosh\left(\dfrac{1}{5}\cosh^{-1}\dfrac{m}{2\,p^{5/2}}\right),m\geq 1 \\ u=2\sqrt{p}\cos\left(\dfrac{1}{5}\cos^{-1}\dfrac{m}{2\,p^{5/2}}\right),-1\leq m\leq 1 \end{cases} \tag{368}$$

> Seventh degree equation and its solution

$$\begin{cases} u^7-7\,p\,u^5+14\,p^2u^3-7\,p^3u-m=0 \\ u=2\sqrt{p}\cosh\left(\dfrac{1}{7}\cosh^{-1}\dfrac{m}{2\,p^{7/2}}\right),m\geq 1 \\ u=2\sqrt{p}\cos\left(\dfrac{1}{7}\cos^{-1}\dfrac{m}{2\,p^{7/2}}\right),-1\leq m\leq 1 \end{cases} \tag{369}$$

Therefore, we do further analysis relating to the triangular coefficients of Chebyshev polynomials [12] and come up with the general equation, and its corresponding general solution:

$$\begin{cases} \left((2n+1)\displaystyle\sum_{k=0}^{n}\left[\dfrac{(-1)^{n-k}(n+k)!\,p^{n-k}u^{2k+1}}{(2k+1)!\,(n-k)!}\right]\right)-m=0 & (370\,a) \\[2ex] u_{k+1}=2\sqrt{p}\cosh\left(\dfrac{1}{2n+1}\cosh^{-1}\left(\dfrac{m}{2\,p^{(2n+1)/2}}\right)+\dfrac{i\,2\,\pi\,k}{2n+1}\right), & \dfrac{m}{2\,p^{(2n+1)/2}}\geq 1 & (370\,b) \\[2ex] =2\sqrt{p}\cosh\left(\dfrac{1}{2n+1}\cosh^{-1}\left(-\dfrac{m}{2\,p^{(2n+1)/2}}\right)+\dfrac{i\,2\,\pi\,k}{2n+1}\right), & \dfrac{m}{2\,p^{(2n+1)/2}}\leq -1 & (370\,b) \\[2ex] =2\sqrt{p}\cos\left(\dfrac{1}{2n+1}\cos^{-1}\left(\dfrac{m}{2\,p^{(2n+1)/2}}\right)+\dfrac{2\,\pi\,k}{2n+1}\right), & -1\leq\dfrac{m}{2\,p^{(2n+1)/2}}\leq 1 & (370\,c) \\[2ex] \text{for } k=0,1,2,\ldots,2n. \end{cases} \tag{370}$$

Furthermore, we can express the solution of equation (370a) in terms of cosht

174

function:

$$u_{k+1}=2\sqrt{p}\cosh_{2n+1,k}\left(\frac{m}{2\,p^{(2n+1)/2}}\right),k=0,1,\ldots,2\,n. \qquad (370\,d)$$

As previously mentioned in #1, by replacing b with -b, we get a similar solution formulas presented in (370) but it has a combined form of hyperbolic sine and its inverse or sine and its inverse. Indeed, we repeat the process from step #1 to #3, we obtain analogous equations with all coefficients being addition signs. The results are shown below:

➢ Nth degree equation and its solution in the form of hyperbolic sine and its inverse

$$\left\{\begin{array}{l} \left((2n+1)\sum_{k=0}^{n}\left[\dfrac{(n+k)!\,p^{n-k}u^{2k+1}}{(2k+1)!\,(n-k)!}\right]\right)+m=0 \qquad\qquad (371\,a) \\[4mm] u_{k+1}=-2\sqrt{p}\sinh\left(\dfrac{1}{2n+1}\sinh^{-1}\left(\dfrac{m}{2\,p^{(2n+1)/2}}\right)\right),\quad -\infty<\dfrac{m}{2\,p^{(2n+1)/2}}<\infty\quad(371\,b) \\[4mm] \text{for } k=0,1,2,\ldots,\;2\,n. \end{array}\right. \qquad (371)$$

➢ And nth degree equation and its solution in the form of sine and its inverse

$$\left\{\begin{array}{l} \left((2n+1)\sum_{k=0}^{n}\left[\dfrac{(-1)^{n-k}(n+k)!\,p^{n-k}u^{2k+1}}{(2k+1)!\,(n-k)!}\right]\right)+m=0 \qquad\qquad (372\,a) \\[4mm] u_{k+1}=(-1)^{n+1}2\sqrt{p}\sin\left(\dfrac{1}{2n+1}\sin^{-1}\left(\dfrac{m}{2\,p^{(2n+1)/2}}\right)\right),-1\le\dfrac{m}{2\,p^{(2n+1)/2}}\le1\quad(372\,b) \\[4mm] \text{for } k=0,1,2,\ldots,\;2\,n. \end{array}\right. \qquad (372)$$

Example. Use the results from (370) to derive the solutions for ninth and eleventh degree equations.

Solution. By substituting n = 4 into (370), we get solution for the ninth degree equation

$$\begin{cases} u^9-9\,pu^7+27\,p^2u^5-30\,p^3u^3+9\,p^4u-m=0 \\ u_{k+1}=2\sqrt{p}\cosh\left(\dfrac{1}{9}\cosh^{-1}\left(\dfrac{m}{2\,p^{9/2}}\right)+\dfrac{i2\pi k}{9}\right), \qquad \dfrac{m}{2\,p^{9/2}}\ge 1 \\ u_{k+1}=2\sqrt{p}\cosh\left(\dfrac{1}{9}\cosh^{-1}\left(\left|\dfrac{m}{2\,p^{9/2}}\right|\right)+\dfrac{i2\pi k}{9}\right), \qquad \dfrac{m}{2\,p^{9/2}}\le -1 \\ u_{k+1}=2\sqrt{p}\cos\left(\dfrac{1}{9}\cos^{-1}\left(\dfrac{m}{2\,p^{9/2}}\right)+\dfrac{2\pi k}{9}\right), \qquad -1\le\dfrac{m}{2\,p^{9/2}}\le 1 \\ \text{for } k=0,1,2,\dots,8. \end{cases} \tag{373}$$

Similarly, by plugging n = 5 in (370), we obtain the eleventh degree equation and its solution

$$\begin{cases} u^{11}-11\,pu^9+44\,p^2u^7-77\,p^3u^5+55\,p^4u^3-11\,p^5u-m=0 \\ u_{k+1}=2\sqrt{p}\cosh t_{11,k}\left(\dfrac{m}{2\,p^{11/2}}\right), \text{ for } k=0,1,2,\dots,10. \end{cases} \tag{374}$$

(Noting that we use the function cosht.)

Example. Use the results from (371) to derive the solutions for ninth and eleventh degree equations.

Solution. By substituting n = 4 and n = 5 into (349) and renaming variable u as x, we obtain the following results:

➢ The ninth degree equation and its solution

$$\begin{cases} x^9+9\,px^7+27\,p^2x^5+30\,p^3x^3+9\,p^4x+m=0 & (375\,a) \\ x_{k+1}=-2\sqrt{p}\sinh\left(\dfrac{1}{9}\sinh^{-1}\left(\dfrac{m}{2\,p^{9/2}}\right)+\dfrac{i2\pi k}{9}\right) & (375\,b) \end{cases} \tag{375}$$

➢ And the eleventh degree equation and its solution

$$\begin{cases} x^{11}+11\,px^9+44\,p^2x^7+77\,p^3x^5+55\,p^4x^3+11\,p^5x+m=0 & (376\,a) \\ x_{k+1}=-2\sqrt{p}\sinh\left(\dfrac{1}{11}\sinh^{-1}\left(\dfrac{m}{2\,p^{11/2}}\right)+\dfrac{i2\pi k}{11}\right) & (376\,b) \end{cases} \tag{376}$$

By using the same method applied to the septic equation in part D of Section IV to construct coefficients a (a≠0), b, and c of equations (375) and (376), which will yield the following results:

➢ Solution to the ninth degree equation (type 1)

176

$$a x^9 + b x^8 + c x^6 + \frac{27 c^2}{100 b} x^4 + \frac{27 c^3}{1000 b^2} x^2 + \frac{9 c^4}{10000 b^3} = 0 \quad (b \neq 0)$$

$$x_{k+1} = -\frac{\sqrt{3} c}{2 \sqrt{10 b} \sinht_{9,k}\left(\frac{9 a \sqrt{3} c}{2 b \sqrt{10 b}}\right)}, \quad k = 0, 1, \ldots, 8. \tag{377}$$

> Solution to the eleventh degree equation

$$a x^{11} + b x^{10} + c x^8 + \frac{7 c^2}{25 b} x^6 + \frac{4 c^3}{125 b^2} x^4 + \frac{c^4}{625 b^3} x^2 + \frac{c^5}{34375 b^4} = 0 \quad (b \neq 0) \tag{378 a}$$

$$x_{k+1} = -\frac{\sqrt{c}}{2 \sqrt{5 b} \sinht_{11,k}\left(\frac{11 a \sqrt{c}}{2 b \sqrt{5 b}}\right)}, \quad k = 0, 1, \ldots, 10. \tag{378 b} \tag{378}$$

By applying (46b), solution formula (378b) to eleventh degree equation (378a) can be converted to the radical form:

$$x_{k+1} = -\frac{\sqrt{c}}{\sqrt{5 b}\left[\left(\frac{11 a \sqrt{c} + \sqrt{121 a^2 c + 20 b^3}}{(20 b^3)^{1/2}}\right)^{1/11} e^{i 2 \pi k/11} - \left(\frac{11 a \sqrt{c} + \sqrt{121 a^2 c + 20 b^3}}{(20 b^3)^{1/2}}\right)^{-1/11} e^{-i 2 \pi k/11}\right]} \tag{378 c}$$

for k = 0, 1, 2, …,10.

I. Solution to Special Form of the Ninth Degree Equation (Type 2)

The special ninth equation is expressed in the following form:

$$a x^9 + c x^7 + d x^6 + e x^5 + \frac{2 c d}{3 a} x^4 + g x^3 + \frac{d e}{3 a} x^2 + \frac{c d^2}{9 a^2} x + \frac{d^3}{27 a^2} = 0 \quad (a \neq 0) \tag{379}$$

In (42), we rename x as u and the coefficients as shown below:

$$a u^3 + a_1 u^2 + a_2 u + a_3 = 0 \quad (a \neq 0) \tag{380 a}$$

$$u_{k+1} = -\frac{a_1}{3 a} + \frac{2}{3 a} \sqrt{a_1^2 - 3 a a_2} \cosh\left(\frac{1}{3} \cosh^{-1}\left(\frac{9 a a_1 a_2 - 2 a_1^3 - 27 a^2 a_3}{2 (a_1^2 - 3 a a_2)^{3/2}}\right) + \frac{i 2 \pi k}{3}\right) \tag{380 b} \tag{380}$$

By replacing u by $(x^2 + d/x)$ and multiplying through equation (380a) by x^3, we obtained

$$\begin{cases} a x^9 + a_1 x^7 + 3 a d x^6 + a_2 x^5 + 2 a_1 d x^4 + \left(a_3 + 3 a d^2\right) x^3 + a_2 d x^2 + a_1 d^2 x + a d^3 = 0 \quad (a \neq 0) & (381\,a) \\ x^2 + \dfrac{d}{x} = u_{k+1} & (381\,b) \end{cases} \quad (381)$$

Rearranging equation (381b) in x, replacing a₁ by c, d by d/(3a) and a₂ by e gives

$$\begin{cases} a x^9 + c x^7 + d x^6 + e x^5 + \dfrac{2 d c}{3 a} x^4 + \left(a_3 + \dfrac{d^2}{3 a}\right) x^3 + \dfrac{d e}{3 a} x^2 + \dfrac{c d^2}{9 a^2} x + \dfrac{d^3}{27 a^2} = 0 \quad (a \neq 0) & (382\,a) \\ 3 a x^3 - 3 a u_{k+1} x + d = 0 & (382\,b) \\ u_{k+1} = -\dfrac{c}{3 a} + \dfrac{2}{3 a} \sqrt{c^2 - 3 a e} \cosh\left(\dfrac{1}{3} \cosh^{-1}\left(\dfrac{9 a c e - 2 c^3 - 27 a^2 a_3}{2\left(c^2 - 3 a e\right)^{3/2}}\right) + \dfrac{i 2 \pi k}{3}\right) & (382\,c) \end{cases} \quad (382)$$

Solving for x from cubic equation (382b) and replacing a₃ with *g – d²/(3a)* gives the desired equation and its solution:

$$\begin{cases} a x^9 + c x^7 + d x^6 + e x^5 + \dfrac{2 c d}{3 a} x^4 + g x^3 + \dfrac{d e}{3 a} x^2 + \dfrac{c d^2}{9 a^2} x + \dfrac{d^3}{27 a^2} = 0 \quad (a \neq 0) & (383 a) \\ x_{j+1} = 2\sqrt{\dfrac{u_{k+1}}{3}} \cosh\left(\dfrac{1}{3} \cosh^{-1}\left(-\dfrac{\sqrt{3} d}{2 a u_{k+1}^{3/2}}\right) + \dfrac{i 2 \pi j}{3}\right) \quad j = 0,1,2 & (383 b) \\ u_{k+1} = -\dfrac{c}{3 a} + \dfrac{2}{3 a} \sqrt{c^2 - 3 a e} \cosh\left(\dfrac{1}{3} \cosh^{-1}\left(\dfrac{9 a d^2 + 9 a c e - 2 c^3 - 27 a^2 g}{2\left(c^2 - 3 a e\right)^{3/2}}\right) + \dfrac{i 2 \pi k}{3}\right) & (383 c) \end{cases} \quad (383)$$

If the inputs of the hyperbolic inverse cosine functions in (383b) and (383c) lie within the interval [-1, 1], then we can replace the functions with the regular cosine and its inverse functions, namely

$$\begin{cases} x_{j+1} = 2\sqrt{\dfrac{u_{k+1}}{3}} \cos\left(\cos^{-1}\left(-\dfrac{\sqrt{3} d}{2 a u^{3/2}}\right) + \dfrac{2 \pi j}{3}\right) \quad j = 0,1,2 & (383\,d) \\ u_{k+1} = -\dfrac{c}{3 a} + \dfrac{2}{3 a} \sqrt{c^2 - 3 a e} \cos\left(\dfrac{1}{3} \cos^{-1}\left(\dfrac{9 a d^2 + 9 a c e - 2 c^3 - 27 a^2 g}{2\left(c^2 - 3 a e\right)^{3/2}}\right) + \dfrac{2 \pi k}{3}\right) & (383\,e) \end{cases}$$

Noting that all solution formulas (383b), (383c), (383b) and (383e) can be rewritten in terms of the cosht function when needed.

For each value of j, formula (383b) gives three roots, corresponding to k = 0, 1 and 2. Therefore, it gives a total of nine roots for equation (383a), corresponding to the combination of j = 0, 1 and 2, as follows:

$$
x_{1,2,3} = \left\{
\begin{array}{l}
2\sqrt{\dfrac{u_1}{3}}\cosh\left(\cosh^{-1}\left(-\dfrac{\sqrt{3}\,d}{2\,a\,u_1^{3/2}}\right)\right) \\[4ex]
2\sqrt{\dfrac{u_2}{3}}\cosh\left(\cosh^{-1}\left(-\dfrac{\sqrt{3}\,d}{2\,a\,u_2^{3/2}}\right)\right) \\[4ex]
2\sqrt{\dfrac{u_3}{3}}\cosh\left(\cosh^{-1}\left(-\dfrac{\sqrt{3}\,d}{2\,a\,u_3^{3/2}}\right)\right)
\end{array}
\right.
\qquad (383\,f)
$$

and

$$
x_{4,5,6} = \left\{
\begin{array}{l}
2\sqrt{\dfrac{u_1}{3}}\cosh\left(\cosh^{-1}\left(-\dfrac{\sqrt{3}\,d}{2\,a\,u_1^{3/2}}\right)+\dfrac{i\,2\pi}{3}\right) \\[4ex]
2\sqrt{\dfrac{u_2}{3}}\cosh\left(\cosh^{-1}\left(-\dfrac{\sqrt{3}\,d}{2\,a\,u_2^{3/2}}\right)+\dfrac{i\,2\pi}{3}\right) \\[4ex]
2\sqrt{\dfrac{u_3}{3}}\cosh\left(\cosh^{-1}\left(-\dfrac{\sqrt{3}\,d}{2\,a\,u_3^{3/2}}\right)+\dfrac{i\,2\pi}{3}\right)
\end{array}
\right.
\qquad (383\,g)
$$

$$
x_{7,8,9} = \left\{
\begin{array}{l}
2\sqrt{\dfrac{u_1}{3}}\cosh\left(\cosh^{-1}\left(-\dfrac{\sqrt{3}\,d}{2\,a\,u_1^{3/2}}\right)+\dfrac{i\,4\pi}{3}\right) \\[4ex]
2\sqrt{\dfrac{u_2}{3}}\cosh\left(\cosh^{-1}\left(-\dfrac{\sqrt{3}\,d}{2\,a\,u_2^{3/2}}\right)+\dfrac{i\,4\pi}{3}\right) \\[4ex]
2\sqrt{\dfrac{u_3}{3}}\cosh\left(\cosh^{-1}\left(-\dfrac{\sqrt{3}\,d}{2\,a\,u_3^{3/2}}\right)+\dfrac{i\,4\pi}{3}\right)
\end{array}
\right.
\qquad (383\,h)
$$

where u_1, u_2 and u_3 are given by formula (383c) or (383e).

K. Solution to the Ninth Degree Equation in Radical Form (Type 2)

Instead of giving a summary section as we have done for others in prior sections, we provide the solution to ninth degree equation (379) in the radical form.

Define

$$D = c^2 - 3ae$$

$$M = 9ad^2 + 9ace - 2c^3 - 27a^2g$$

Then solution formula (383c) becomes

$$u_{k+1} = -\frac{c}{3a} + \frac{2}{3a}\sqrt{D}\cosh\left(\frac{1}{3}\cosh^{-1}\left(\frac{M}{2D^{3/2}}\right) + \frac{i2\pi k}{3}\right), \quad k=0,1,2. \tag{384}$$

By applying (46a) to both (384) and (383b), which can be converted to the following radical forms:

$$u_{k+1} = -\frac{c}{3a} + \frac{1}{3a}\sqrt{D}\left[\left(\frac{M+\sqrt{M^2-4D^3}}{2D^{3/2}}\right)^{1/3} e^{i2\pi k/3} + \left(\frac{M+\sqrt{M^2-4D^3}}{2D^{3/2}}\right)^{-1/3} e^{-i2\pi k/3}\right]$$

$$u_{k+1} = -\frac{c}{3a} + \frac{1}{3a}\sqrt{D}\left[\left(\frac{M+\sqrt{M^2-4D^3}}{2D^{3/2}}\right)^{1/3} e^{i2\pi k/3} + \left(\frac{M+\sqrt{M^2-4D^3}}{2D^{3/2}}\right)^{-1/3} e^{-i2\pi k/3}\right]$$

$$= -\frac{c}{3a} + \frac{1}{3a}\left[\left(\frac{1}{2}\left(M+\sqrt{M^2-4D^3}\right)\right)^{1/3} e^{i2\pi k/3} + D\left(\frac{1}{2}\left(M+\sqrt{M^2-4D^3}\right)\right)^{-1/3} e^{-i2\pi k/3}\right] \tag{385}$$

for k = 0, 1, 2.

and

$$x_{j+1} = 2\sqrt{\frac{u_{k+1}}{3}}\frac{1}{2}\left[\left(-\frac{\sqrt{3}d}{2au_{k+1}^{3/2}} + \sqrt{\frac{3d^2}{4a^2u_{k+1}^3} - 1}\right)^{1/3} e^{i2\pi j/3} + \left(-\frac{\sqrt{3}d}{2au_{k+1}^{3/2}} + \sqrt{\frac{3d^2}{4a^2u_{k+1}^3} - 1}\right)^{-1/3} e^{-i2\pi j/3}\right]$$

$$= \frac{1}{\sqrt{3}}\left[\left(\frac{1}{2a}\left(-\sqrt{3}d + \sqrt{3d^2-4a^2u_{k+1}^3}\right)\right)^{1/3} e^{i2\pi j/3} + u_{k+1}\left(\frac{1}{2a}\left(-\sqrt{3}d + \sqrt{3d^2-4a^2u_{k+1}^3}\right)\right)^{-1/3} e^{-i2\pi j/3}\right] \tag{386}$$

for j = 0, 1, 2.

Hence, by combining the results from (385) and (386), the roots of the ninth degree equation (379) can be expressed explicitly in radical form:

$$x_1 = \frac{1}{\sqrt{3}}\left[\left(\frac{1}{2a}\left(-\sqrt{3}d + \sqrt{3d^2-4a^2u_1^3}\right)\right)^{1/3} + u_1\left(\frac{1}{2a}\left(-\sqrt{3}d + \sqrt{3d^2-4a^2u_1^3}\right)\right)^{-1/3}\right]$$

$$x_2 = \frac{1}{\sqrt{3}}\left[\left(\frac{1}{2a}\left(-\sqrt{3}d + \sqrt{3d^2-4a^2u_1^3}\right)\right)^{1/3} e^{i2\pi/3} + u_1\left(\frac{1}{2a}\left(-\sqrt{3}d + \sqrt{3d^2-4a^2u_1^3}\right)\right)^{-1/3} e^{-i2\pi/3}\right]$$

$$x_3 = \frac{1}{\sqrt{3}}\left[\left(\frac{1}{2a}\left(-\sqrt{3}d + \sqrt{3d^2-4a^2u_1^3}\right)\right)^{1/3} e^{i4\pi/3} + u_1\left(\frac{1}{2a}\left(-\sqrt{3}d + \sqrt{3d^2-4a^2u_1^3}\right)\right)^{-1/3} e^{-i4\pi/3}\right]$$

$$x_4 = \frac{1}{\sqrt{3}}\left[\left(\frac{1}{2a}\left(-\sqrt{3}\,d + \sqrt{3\,d^2 - 4\,a^2\,u_2^3}\right)\right)^{1/3} + u_2\left(\frac{1}{2a}\left(-\sqrt{3}\,d + \sqrt{3\,d^2 - 4\,a^2\,u_2^3}\right)\right)^{-1/3}\right]$$

$$x_5 = \frac{1}{\sqrt{3}}\left[\left(\frac{1}{2a}\left(-\sqrt{3}\,d + \sqrt{3\,d^2 - 4\,a^2\,u_2^3}\right)\right)^{1/3}e^{i2\pi/3} + u_2\left(\frac{1}{2a}\left(-\sqrt{3}\,d + \sqrt{3\,d^2 - 4\,a^2\,u_2^3}\right)\right)^{-1/3}e^{-i2\pi/3}\right]$$

$$x_6 = \frac{1}{\sqrt{3}}\left[\left(\frac{1}{2a}\left(-\sqrt{3}\,d + \sqrt{3\,d^2 - 4\,a^2\,u_2^3}\right)\right)^{1/3}e^{i4\pi/3} + u_2\left(\frac{1}{2a}\left(-\sqrt{3}\,d + \sqrt{3\,d^2 - 4\,a^2\,u_2^3}\right)\right)^{-1/3}e^{-i4\pi/3}\right]$$

$$x_7 = \frac{1}{\sqrt{3}}\left[\left(\frac{1}{2a}\left(-\sqrt{3}\,d + \sqrt{3\,d^2 - 4\,a^2\,u_3^3}\right)\right)^{1/3} + u_3\left(\frac{1}{2a}\left(-\sqrt{3}\,d + \sqrt{3\,d^2 - 4\,a^2\,u_3^3}\right)\right)^{-1/3}\right]$$

$$x_8 = \frac{1}{\sqrt{3}}\left[\left(\frac{1}{2a}\left(-\sqrt{3}\,d + \sqrt{3\,d^2 - 4\,a^2\,u_3^3}\right)\right)^{1/3}e^{i2\pi/3} + u_3\left(\frac{1}{2a}\left(-\sqrt{3}\,d + \sqrt{3\,d^2 - 4\,a^2\,u_3^3}\right)\right)^{-1/3}e^{-i2\pi/3}\right]$$

$$x_9 = \frac{1}{\sqrt{3}}\left[\left(\frac{1}{2a}\left(-\sqrt{3}\,d + \sqrt{3\,d^2 - 4\,a^2\,u_3^3}\right)\right)^{1/3}e^{i4\pi/3} + u_3\left(\frac{1}{2a}\left(-\sqrt{3}\,d + \sqrt{3\,d^2 - 4\,a^2\,u_3^3}\right)\right)^{-1/3}e^{-i4\pi/3}\right]$$

where u_1, u_2 and u_3 are

$$u_1 = -\frac{c}{3a} + \frac{1}{3a}\left[\left(\frac{1}{2}\left(M + \sqrt{M^2 - 4D^3}\right)\right)^{1/3} + D\left(\frac{1}{2}\left(M + \sqrt{M^2 - 4D^3}\right)\right)^{-1/3}\right]$$

$$u_2 = -\frac{c}{3a} + \frac{1}{3a}\left[\left(\frac{1}{2}\left(M + \sqrt{M^2 - 4D^3}\right)\right)^{1/3}e^{i2\pi/3} + D\left(\frac{1}{2}\left(M + \sqrt{M^2 - 4D^3}\right)\right)^{-1/3}e^{-i2\pi/3}\right]$$

$$u_3 = -\frac{c}{3a} + \frac{1}{3a}\left[\left(\frac{1}{2}\left(M + \sqrt{M^2 - 4D^3}\right)\right)^{1/3}e^{i4\pi/3} + D\left(\frac{1}{2}\left(M + \sqrt{M^2 - 4D^3}\right)\right)^{-1/3}e^{-i4\pi/3}\right]$$

Notice that in equation (379), the coefficient b in the x^8 term is absent. We can have this coefficient in the equation by substituting x with x + b/(9a) and then repeat the process of generating other coefficients, but we have not shown the works here due to the large terms generated during the process.

Appendix 2-A: Miscellaneous Proofs of Various Identities and Formulas

[2-A-0]

The following trigonometric identities are well-established and commonly used in mathematics.

$\sin(ix)=i\sinh(x)$	$\tan(ix)=i\tanh(x)$	$\sinh(ix)=i\sin(x)$	$\tanh(ix)=\tan(x)$
$\cos(ix)=\cos(x)$	$\cot(ix)=i\coth(x)$	$\cosh(ix)=\cos(x)$	$\coth(ix)=\cot(x)$
$\sin^{-1}(-x)=-\sin^{-1}(x)$	$\cos^{-1}(-x)=\pi-\cos^{-1}(x)$	$\cosh^{-1}(-x)=\pi i-\cosh^{-1}(x)$	$\cos^{-1}(-x)=\pi-\cos^{-1}(x)$
$\sinh^{-1}(-x)=-\sinh^{-1}(x)$	$\cosh^{-1}(-x)=\pi i-\cosh^{-1}(x)$	$\coth^{-1}(-x)=-\coth^{-1}(x)$	$\cot^{-1}(-x)=-\cot^{-1}(x)$
$\tan^{-1}(ix)=i\tanh^{-1}(x)$	$\cot^{-1}(ix)=-i\coth^{-1}(x)$	$i\cos^{-1}(x)=\cosh^{-1}(x)$	$\sin^{-1}(ix)=i\sinh^{-1}(x)$
$\tanh^{-1}(ix)=i\tan^{-1}(x)$	$\coth^{-1}(iu)=-i\cot^{-1}(x)$	$i\cosh^{-1}(x)=\cos^{-1}(x)$	$\sinh^{-1}(ix) = i\sin^{-1}(x)$

$\sinh(x+i2k\pi)=\sinh(x)$, $k\in\mathbb{Z}$ $\qquad$ $\tanh(x+i\pi k)=\tanh(x)$, $k\in\mathbb{Z}$

$\cosh(x+i2k\pi)=\cosh(x)$, $k\in\mathbb{Z}$ $\qquad$ $\coth(x+i\pi k)=\coth(x)$, $k\in\mathbb{Z}$

[2-A-1]

- $\cos(2x)=2\cos^2 x-1$ (Double angle identity for cosine)
  $\qquad =1-2\sin^2 x$

- $\cos(3x)=4\cos^3 x-3\cos x$ (Triple angle identity for cosine)

- $\cos(4x)=8\cos^4 x-8\cos^2 x+1$ (Quadruple angle identity for cosine)

- $\cos(5x)=16\cos^5 x-20\cos^3 x+5\cos x$ (Quintuple angle identity for cosine)

- $\cos(6x)=32\cos^6 x-48\cos^4 x+18\cos^2 x-1$ (Sextuple angle identity for cosine)

- $\cos(7x)=64\cos^7 x-112\cos^5 x+56\cos^3 x-7\cos x$ (Septic angle identity for cosine)

See more>> http://www.seriesmathstudy.com/sms/multipleanglecosine.

[2-A-2]

We show the solution of the composition of the cosine function with its inverse. Since the cosine function is periodic with a period of 2π, it can be expressed as

$$\cos(x)=\cos(x\pm2\pi k),\ \forall\,x,k\in\mathbb{Z}.$$

Hence, to solve for

$$\cos\left(n\cos^{-1}(x)\right)=m,\quad -1\le x\le1,\text{m and n}\in\mathbb{R},$$

we can manipulate it to get x through the following steps:

$$\Leftrightarrow \cos\left(n\cos^{-1}(x)\pm2\pi k\right)=m,\qquad -1\le x\le1,k\in\mathbb{Z}\text{, m and n}\in\mathbb{R}.$$

$$\Leftrightarrow n\cos^{-1}(x)\pm2\pi k=\cos^{-1}(m),$$

$$\Leftrightarrow \cos^{-1}(x)\pm2\pi k=\frac{1}{n}\cos^{-1}(m),$$

$$\Leftrightarrow \cos^{-1}(x)=\frac{1}{n}\cos^{-1}(m)\mp2\pi k,$$

$$\Leftrightarrow x=\cos\left(\frac{1}{n}\cos^{-1}(m)\mp\frac{2\pi k}{n}\right),\qquad -1\le x\le1,k\in\mathbb{Z}\text{, and }m,n\in\mathbb{R}.$$

The same approach can be applied to other trigonometric and hyperbolic trigonometric functions provided that an equivalent period of each trigonometric function is specified.

[2-A-3]

- $$\cosh(2x)=2\cosh^2x-1$$
  $$=1+2\sinh^2x$$
  (Double angle identity for hyperbolic cosine)

- $$\cosh(3x)=4\cosh^3x-3\cosh x$$
  (Triple angle identity for hyperbolic cosine)

- $$\cosh(4x)=8\cosh^4x-8\cosh^2x+1$$
  (Quadruple angle identity for hyperbolic cosine)

- $$\cosh(5x)=16\cosh^5x-20\cosh^3x+5\cosh x$$
  (Quintuple angle identity for hyperbolic cosine)

- $$\cosh(6x)=32\cosh^6x-48\cosh^4x+18\cosh^2x-1$$
  (Sextuple angle identity for hyperbolic cosine)

- $$\cos(7x)=64\cos^7x-112\cos^5x+56\cos^3x-7\cos x$$

(Septic angle hyperbolic cosine identity)

See more>> http://seriesmathstudy.com/sms/multipleanglehyperboliccosine.

[2-A-4]

In [2-A-1] substituting x by cos⁻¹(x) into the quadruple angle cosine identity, we have

$$\cos(4x) = 8\cos^4 x - 8\cos^2 x + 1.$$

Applying the formula cos⁻¹(x) = π/2 – sin⁻¹(x) gives

$$\cos\left(2\pi - 4\sin^{-1}(x)\right) = 8x^4 - 8x^2 + 1$$

$$\Rightarrow \cos\left(-4\sin^{-1}(x)\right) = 8x^4 - 8x^2 + 1$$

$$\Rightarrow \cos\left(4\sin^{-1}(x)\right) = 8x^4 - 8x^2 + 1.$$

[2-A-5]

We apply double angle identities for cosine and hyperbolic cosine to express solutions (20b) and (23b) in terms of radicals as follows:

- Using (20b) with k = 0 and k = 1 gives

$$x_{1,2} = -\frac{b}{2a}\left(1 \pm \cos\left[\frac{1}{2}\cos^{-1}\left(\frac{b^2 - 8ac}{b^2}\right)\right]\right) = \begin{cases} -\dfrac{b}{a}\cos^2\left(\dfrac{1}{4}\cos^{-1}\left(\dfrac{b^2-8ac}{b^2}\right)\right) & (a1) \\[4mm] -\dfrac{b}{a}\sin^2\left(\dfrac{1}{4}\cos^{-1}\left(\dfrac{b^2-8ac}{b^2}\right)\right) & (a2) \end{cases}$$

- Applying formulas (1.5) and (1.4.4) gives

$$x_{1,2} = -\frac{b}{2a}\left(1 \pm \cos\left[\frac{1}{2}\cos^{-1}\left(\frac{b^2 - 8ac}{b^2}\right)\right]\right) = \begin{cases} -\dfrac{b}{4a}\left(2 + \sqrt{2 + 2\dfrac{b^2-8ac}{b^2}}\right) \\[4mm] -\dfrac{b}{4a}\left(2 - \sqrt{2 + 2\dfrac{b^2-8ac}{b^2}}\right) \end{cases}$$

$$= \begin{cases} -\dfrac{1}{2a}\left(b + \sqrt{b^2 - 4ac}\right) & (a3) \\[4mm] -\dfrac{1}{2a}\left(b - \sqrt{b^2 - 4ac}\right). & (a4) \end{cases}$$

- Using (23b) with k = 0 and k = 1 gives

$$x_{1,2}=-\frac{b}{2a}\left(1\pm\cosh\left[\frac{1}{2}\cosh^{-1}\left(\frac{b^2-8ac}{b^2}\right)\right]\right)=\begin{cases}-\dfrac{b}{a}\cosh^2\left(\dfrac{1}{4}\cosh^{-1}\left(\dfrac{b^2-8ac}{b^2}\right)\right) & (a5)\\[2em]\dfrac{b}{a}\sinh^2\left(\dfrac{1}{4}\cosh^{-1}\left(\dfrac{b^2-8ac}{b^2}\right)\right) & (a6)\end{cases}$$

- By applying (1.2.3) and (1.5.3), which gives

$$=\begin{cases}-\dfrac{b}{4a}\left(2+\sqrt{2+2\dfrac{b^2-8ac}{b^2}}\right)\\[2em]\dfrac{b}{4a}\left(\sqrt{2+2\dfrac{b^2-8ac}{b^2}}-2\right)\end{cases}$$

$$=\begin{cases}-\dfrac{1}{2a}\left(b+\sqrt{b^2-4ac}\right) & (a7)\\[1.5em]-\dfrac{1}{2a}\left(b-\sqrt{b^2-4ac}\right). & (a8)\end{cases}$$

[2-A-6]

We use (40b),

$$x_{k+1}=-\frac{b}{3a}+\frac{2}{3a}\sqrt{b^2-3ac}\cos\left[\frac{1}{3}\cos^{-1}(T)+\frac{2\pi k}{3}\right]\text{ for }-1\leq T\leq1,$$

where

$$T=\frac{9abc-2b^3-27a^2d}{2(b^2-3ac)^{3/2}},$$

to derive (42b) as follows:

If T≥1, by applying $\cos^{-1}(u)=\dfrac{1}{i}\ln\left(u+i\sqrt{1-u^2}\right)$ to formula (40b), which is written as

$$x_{k+1}=-\frac{b}{3a}+\frac{2}{3a}\sqrt{b^2-3ac}\cos\left[\frac{1}{3}\left(\frac{1}{i}\ln\left(T-\sqrt{T^2-1}\right)\right)+\frac{2\pi k}{3}\right]$$

$$=-\frac{b}{3a}+\frac{2}{3a}\sqrt{b^2-3ac}\cos\left[\frac{1}{3}\left(\frac{1}{i}\ln\left(\frac{1}{T+\sqrt{T^2-1}}\right)+\frac{2\pi k}{3}\right)\right]$$

$$=-\frac{b}{3a}+\frac{2}{3a}\sqrt{b^2-3ac}\cos\left[\frac{1}{3}\left(-\frac{1}{i}\ln\left(T+\sqrt{T^2-1}\right)+\frac{2\pi k}{3}\right)\right]$$

Substituting $\ln(T+\sqrt{T^2-1})=\cosh^{-1}(T)$ gives

$$=-\frac{b}{3a}+\frac{2}{3a}\sqrt{b^2-3ac}\cos\left(\frac{i}{3}\cosh^{-1}(T)+\frac{2\pi k}{3}\right)$$

Factoring out i from the expression in parentheses gives

$$=-\frac{b}{3a}+\frac{2}{3a}\sqrt{b^2-3ac}\cos\left[i\left(\frac{1}{3}\cosh^{-1}(T)-\frac{i2\pi k}{3}\right)\right]$$

Use cos(ix) = cosh(x). Since the value of k can be 0, ±1, ±2, by re-indexing the value of k in the cosine function, the roots x_{k+1} remain unchanged without loss of generality:

$$=-\frac{b}{3a}+\frac{2}{3a}\sqrt{b^2-3ac}\cosh\left[\frac{1}{3}\cosh^{-1}(T)+\frac{i2\pi k}{3}\right] \quad \text{for } |T|\geq 1.$$

[2-A-7]

1. $\displaystyle\lim_{x\to 0}\sqrt{x}\cosh\left(\frac{1}{2}\ln(x)\right)=\lim_{x\to 0}\sqrt{x}\cosh\left(\ln\left(\sqrt{x}\right)\right)$

$$=\lim_{x\to 0}\frac{\sqrt{x}}{2}\left(\sqrt{x}+\frac{1}{\sqrt{x}}\right)$$

$$=\lim_{x\to 0}\frac{1}{2}(x+1)$$

$$=\frac{1}{2}.$$

2. $\displaystyle\lim_{x\to 0}\sqrt{x}\sinh\left(\frac{1}{2}\ln(x)\right)=\lim_{x\to 0}\sqrt{x}\sinh\left(\ln\left(\sqrt{x}\right)\right)$

$$=\lim_{x\to 0}\frac{\sqrt{x}}{2}\left(\sqrt{x}-\frac{1}{\sqrt{x}}\right)$$

$$=\lim_{x\to 0}\frac{1}{2}(x-1)$$

$$=-\frac{1}{2}.$$

[2-A-8]
Triple Angle Identity for Sine
By taking the derivative of both sides of triple angle identity for cosine in [2-A-1], we have:

$$-3\sin(3x)=-12\sin x\cos^2 x+3\sin x$$
$$\Rightarrow \sin(3x)=4\sin x\cos^2 x-\sin x$$
$$\Rightarrow \sin(3x)=4\sin x(1-\sin^2 x)-\sin x$$
$$\Rightarrow \sin(3x)=-4\sin^3 x+3\sin x \quad \text{(Triple angle identity for sine)}$$

Triple Angle Identity for Hyperbolic Sine

In [2-A-8], replace x by ix into the triple angle identity for sine. We find that

$$\sin(3ix)=-4\sin^3 ix+3\sin ix$$
$$\Rightarrow i\sinh(3x)=4i\sinh^3 x+3i\sinh x$$
$$\Rightarrow \sinh(3x)=4\sinh^3 x+3\sinh x \quad \text{(Triple angle identity for hyperbolic sine)}$$

[2-A-9]
Triple Angle Identity for Tangent

The triple angle identity can be derived as follows:

$$\tan(3x)=\frac{\sin(3x)}{\cos(3x)}$$
$$=\frac{3\sin x-4\sin^3 x}{4\cos^3 x-3\cos x}$$
$$=\frac{\dfrac{3\tan x}{\cos^2 x}-4\tan^3 x}{4-\dfrac{3}{\cos^2 x}}$$
$$=\frac{3\tan x(\tan^2 x+1)-4\tan^3 x}{4-3(\tan^2 x+1)}$$
$$=\frac{3\tan x-\tan^3 x}{1-3\tan^2 x} \quad \text{(triple angle identity for tangent)}$$

See more>> http://seriesmathstudy.com/sms/multiple_angle_tangent

Triple Angle Identity for Hyperbolic Tangent

In [2-A-9] replace x by ix in triple angle identity for tangent to get

$$\tanh(3x)=\frac{3\tanh x+\tanh^3 x}{1+3\tanh^2 x}.$$

See more>> http://www.seriesmathstudy.com/sms/multiple_angle_hyper_tangent.

[2-A-10]

Simplifying the formula:

$$x_{k+1} = -\frac{c}{4b} - \frac{2b^2 - 3ac}{4ab} \tanh\left(\frac{1}{3}\tanh^{-1}(1) + \frac{\pi i k}{3}\right), \quad k = 0, 1, 2.$$

We take a look at this expression,

$$\tanh\left(\frac{1}{3}\tanh^{-1} u\right) = \frac{\sqrt[6]{\frac{1+u}{1-u}} - \sqrt[6]{\frac{1-u}{1+u}}}{\sqrt[6]{\frac{1+u}{1-u}} + \sqrt[6]{\frac{1-u}{1+u}}}$$

$$= \frac{\sqrt[3]{1+u} - \sqrt[3]{1-u}}{\sqrt[3]{1+u} + \sqrt[3]{1-u}}.$$

Then taking the limit as u tends to 1, we find that

$$\lim_{u \to 1} \tanh\left(\frac{1}{3}\tanh^{-1}(u) + \frac{\pi i k}{3}\right) = \lim_{u \to 1} \frac{\sqrt[3]{1+u}\, e^{i\pi k/3} - \sqrt[3]{1-u}\, e^{-i\pi k/3}}{\sqrt[3]{1+u}\, e^{i\pi k/3} + \sqrt[3]{1-u}\, e^{-i\pi k/3}}$$

$$= \lim_{u \to 1} \frac{\sqrt[3]{1+u}\, e^{i\pi k/3}}{\sqrt[3]{1+u}\, e^{i\pi k/3}} = 1.$$

Hence, the given formula is rewritten as

$$x_{1,2,3} = -\frac{c}{4b} - \frac{2b^2 - 3ac}{4ab}$$

$$= \frac{ac - b^2}{2ab}.$$

[2-A-11]

Consider taking the limit of the expression, $Au\cosh\left(\frac{1}{3}\cosh^{-1}\frac{B}{u^3}\right)$, as u tends to 0:

$$\lim_{u \to 0} A\, u \cosh\left(\frac{1}{3}\cosh^{-1}\frac{B}{u^3}\right) = \lim_{u \to 0} A\, u \cosh\left(\frac{1}{3}\ln\left(\frac{B}{u^3} + \sqrt{\frac{B^2}{u^6} - 1}\right)\right)$$

$$= \lim_{u \to 0} A\, u \cosh\left(\frac{1}{3}\ln\left(\frac{B}{u^3} + \frac{1}{u^3}\sqrt{B^2 - u^6}\right)\right)$$

$$= \lim_{u \to 0} A\, u \cosh\left(\frac{1}{3}\ln\left(B + \sqrt{B^2 - u^6}\right) - \ln u\right)$$

$$= \lim_{u \to 0} A\, u \cosh\left(\frac{1}{3}\ln\left(B + \sqrt{B^2 - u^6}\right)\right)\cosh(\ln u)$$

$$-A\,u\,\sinh\left(\frac{1}{3}\ln\left(B+\sqrt{B^2-u^6}\right)\right)\sinh\left(\ln u\right)$$

(Using results [2-A-7]: $\lim_{x\to 0} x\cosh\left(\ln(x)\right)=\frac{1}{2}$, and $\lim_{x\to 0} x\sinh\left(\ln(x)\right)=-\frac{1}{2}$.)

$$=\frac{A}{2}\cosh\left(\frac{1}{3}\ln(2B)\right)+\frac{A}{2}\sinh\left(\frac{1}{3}\ln(2B)\right)$$

$$=\frac{A}{2}\left[\cosh\left(\frac{1}{3}\ln(2B)\right)+\sinh\left(\frac{1}{3}\ln(2B)\right)\right]$$

$$=\frac{A}{2}\exp\left(\frac{1}{3}\ln(2B)+\frac{2\pi i k}{3}\right)$$

$$=\frac{A\sqrt[3]{2B}}{2}\exp\left(\frac{2\pi i k}{3}\right) \tag{45}$$

Substituting A and B into (45) yields the desired result (43).

[2-A-12]

A. Solution to Depressed Quartic Equation

The depressed quartic equation can be written in the form of

$$a x^4 + b x^2 + c x + d = 0 \quad (a\neq 0), \tag{1}$$

where a, b, c and d are real numbers. The x^3 term is omitted.

The quicker approach involves setting b to zero in either expressions (84) or (85) as depicted in the step 16 of Section III. Or we substitute x with x-b/(4a) and redo the process of constructing coefficients which leads a similar result. The alternative method assumes that neither (84) nor (85) has been derived yet, we return to Step 7 of Section III and begin process of deriving the depressed quartic equation. For more convenience, we rename all coefficients of (76) of Section III to all capital letters as follows.

$$\begin{cases} Ax^4+Bx^3+Cx^2+Dx+\dfrac{AD^2}{B^2}=0 \quad (A\neq 0) \\[2em] x=-\dfrac{B}{4A}\left(1\pm\cos\left[\dfrac{1}{2}\cos^{-1}\left(\dfrac{B^3+16A^2D-8ABC}{B^3}\right)\right]\right)\pm\dfrac{1}{4A}\sqrt{B^2\left(1\pm\cos\left[\dfrac{1}{2}\cos^{-1}\left(\dfrac{B^3+16A^2D-8ABC}{B^3}\right)\right]\right)^2-\dfrac{16A^2D}{B}}. \end{cases} \tag{2}$$

1. By substituting x with x-B/(4A) into (2), which vanishes the x^3 term, resulting in a depressed quartic.

189

$$\begin{cases} Ax^4 + \left(C - \dfrac{3B^2}{8A}\right)x^2 + \left(A - \dfrac{B^3}{8A^2} - \dfrac{BC}{2A}\right)x + \dfrac{AD^2}{B^2} - \dfrac{3B^4}{256A^3} + \dfrac{B^2C}{16A^2} - BD = 0 \quad (A \neq 0) \\[4mm] x = \mp \dfrac{B}{4A}\cos\left[\dfrac{1}{2}\cos^{-1}\left(\dfrac{B^3 + 16A^2D - 8ABC}{B^3}\right)\right] \pm \dfrac{1}{4A}\sqrt{B^2\left(1 \pm \cos\left[\dfrac{1}{2}\cos^{-1}\left(\dfrac{B^3 + 16A^2D - 8ABC}{B^3}\right)\right]\right)^2 - \dfrac{16A^2D}{B}} \end{cases} \quad (3)$$

2. Replacing A with a and C with $b + \dfrac{3B^2}{8A}$ in (3) gives

$$\begin{cases} ax^4 + bx^2 + \left(D - \dfrac{Bb}{2a} - \dfrac{B^3}{16A^2}\right)x - \dfrac{BD}{4a} + \dfrac{D^2a}{B^2} + \dfrac{B^2b}{16a^2} + \dfrac{3B^4}{256a^3} = 0 \quad (a \neq 0) \\[4mm] x = -\dfrac{b}{4a}\left(1 + \cos\left[\dfrac{1}{2}\cos^{-1}\left(\dfrac{b^3 + 16a^2d - 8abc}{b^3}\right)\right]\right) \pm \dfrac{1}{4a}\sqrt{b^2\left(1 \pm \cos\left[\dfrac{1}{2}\cos^{-1}\left(\dfrac{b^3 + 16a^2d - 8abc}{b^3}\right)\right]\right)^2 - \dfrac{16a^2d}{b}} \end{cases} \quad (4)$$

3. Replacing D with $c + \dfrac{Bb}{2a} + \dfrac{B^3}{16a^2}$ and B with ρ into (4) gives

$$\begin{cases} ax^4 + bx^2 + cx - \dfrac{\rho c}{8a} + \dfrac{ac^2}{\rho^2} + \dfrac{bc}{\rho} + \dfrac{b^2}{4a} = 0 & (5a) \\[4mm] x = \mp \dfrac{\rho}{4a}\cos\left[\dfrac{1}{2}\cos^{-1}\left(\dfrac{16a^2c - \rho^3}{\rho^3}\right)\right] \pm \dfrac{1}{4a}\sqrt{\rho^2\left(1 \pm \cos\left[\dfrac{1}{2}\cos^{-1}\left(\dfrac{16a^2c - \rho^3}{\rho^3}\right)\right]\right)^2 - \dfrac{8ab\rho + \rho^3 + 16a^2c}{\rho}} & (5b) \end{cases} \quad (5)$$

4. Setting

$$-\dfrac{\rho c}{8a} + \dfrac{ac^2}{\rho^2} + \dfrac{bc}{\rho} + \dfrac{b^2}{4a} = d \quad (*)$$

5. Multiplying through both sides of (*) by $8a\rho^2$ and isolating ρ produces a cubic equation:

$$c\rho^3 + \left(8ad - 2b^2\right)\rho^2 - 8abc\rho - 8a^2c^2 = 0.$$

6. Using cubic formula (40b) to solve for ρ gives:

$$\rho = \dfrac{2\left(b^2 - 4ad\right)}{3c} + \dfrac{4\sqrt{b^4 + 16a^2d^2 + 6abc^2 - 8ab^2d}}{3c}\cos\left(\dfrac{1}{3}\cos^{-1}\dfrac{2b^6 + 18ab^3c^2 + 27a^2c^4 + 96a^2b^2d^2 - 24ab^4d - 72a^2bc^2d - 128a^3d^3}{2\left(b^4 + 16a^2d^2 + 6abc^2 - 8ab^2d\right)^{3/2}}\right) \quad (6)$$

Therefore, the solution to the depressed quartic is expressed in terms of cosine and its inverse after substituting expression (*) for d in (5a):

$$
\begin{cases}
ax^4+bx^2+cx+d=0 \quad (a\neq 0) & (7a)\\[2mm]
x_{1,2,3,4}=\mp\dfrac{\rho}{4a}\cos\left[\dfrac{1}{2}\cos^{-1}\left(\dfrac{16a^2c-\rho^3}{\rho^3}\right)\right]\pm\dfrac{1}{4a}\sqrt{\rho^2\left(1\pm\cos\left[\dfrac{1}{2}\cos^{-1}\left(\dfrac{16a^2c-\rho^3}{\rho^3}\right)\right]\right)^2-\dfrac{\rho^3+16a^2c+8ab\rho}{\rho}} & (7b)\\[2mm]
\text{where}\\[1mm]
\rho=\dfrac{2(b^2-4ad)}{3c}+\dfrac{4\sqrt{b^4+16a^2d^2+6abc^2-8ab^2d}}{3c}\cos\left(\dfrac{1}{3}\cos^{-1}\dfrac{2b^6+18ab^3c^2+27a^2c^4+96a^2b^2d^2-24ab^4d-72a^2bc^2d-128a^3d^3}{2\left(b^4+16a^2d^2+6abc^2-8ab^2d\right)^{3/2}}\right)
\end{cases}
$$

(7)

and the same solution expressed using hyperbolic cosine and its inverse:

$$
\begin{cases}
ax^4+bx^2+cx+d=0 \quad (a\neq 0) & (8a)\\[2mm]
x_{1,2,3,4}=\mp\dfrac{\rho}{4a}\cosh\left[\dfrac{1}{2}\cosh^{-1}\left(\dfrac{16a^2c-\rho^3}{\rho^3}\right)\right]\pm\dfrac{1}{4a}\sqrt{\rho^2\left(1\pm\cosh\left[\dfrac{1}{2}\cosh^{-1}\left(\dfrac{16a^2c-\rho^3}{\rho^3}\right)\right]\right)^2-\dfrac{\rho^3+16a^2c+8ab\rho}{\rho}} & (8b)\\[2mm]
\text{where}\\[1mm]
\rho=\dfrac{2(b^2-4ad)}{3c}+\dfrac{4\sqrt{b^4+16a^2d^2+6abc^2-8ab^2d}}{3c}\cosh\left(\dfrac{1}{3}\cosh^{-1}\dfrac{2b^6+18ab^3c^2+27a^2c^4+96a^2b^2d^2-24ab^4d-72a^2bc^2d-128a^3d^3}{2\left(b^4+16a^2d^2+6abc^2-8ab^2d\right)^{3/2}}\right)
\end{cases}
$$

(8)

We note that ρ is determined by cubic formula (6) that may involve cosine and its inverse or hyperbolic cosine and its inverse, detailed in Section II. In addition, we aim to present solution(s) in various formats and use different approaches to enhance readers' comprehensive understanding of the method.

B. Transforming Depressed Quartic Equation and Its Solution into General Form

We proceed by transforming the depressed quartic (7) or (8) and its solution into the general quartic. To make the process more convenient, we will rename the coefficients a, b, c, and d of equation (7) to A, C, and D, respectively. Hence, the depressed quartic (1) becomes

$$\left\{ ax^4 + Bx^2 + Cx + D = 0 \quad (a \neq 0) \right. \tag{9a}$$

$$x = \mp \frac{\rho}{4a} \cos\left[\frac{1}{2}\cos^{-1}\left(\frac{16a^2C - \rho^3}{\rho^3}\right)\right] \pm \frac{1}{4a}\sqrt{\rho^2\left(1 \pm \cos\left[\frac{1}{2}\cos^{-1}\left(\frac{16a^2C - \rho^3}{\rho^3}\right)\right]\right)^2 - \frac{\rho^3 + 16a^2C + 8aB\rho}{\rho}} \tag{9b}$$

where $\quad$ (9)

$$\rho = \frac{2(B^2 - 4aD)}{3C} + \frac{4\sqrt{B^4 + 16a^2D^2 + 6aBC^2 - 8aB^2D}}{3C} *$$

$$\cos\left(\frac{1}{3}\cos^{-1}\frac{2B^6 + 18aB^3C^2 + 27a^2C^4 + 96a^2B^2D^2 - 24aB^4D - 72a^2BC^2D - 128a^3D^3}{2(B^4 + 16a^2D^2 + 6aBC^2 - 8aBb^2D)^{3/2}}\right) \tag{9c}$$

1. By replacing x with x+b/(4a) in (9a) and (9b), which gives

$$\left\{ ax^4 + bx^3 + \left(B + \frac{3b^2}{8a}\right)x^2 + \left(C + \frac{Bb}{2a} + \frac{b^3}{16a^2}\right)x + D + \frac{Bb^2}{16a^2} + \frac{Cb}{4a} + \frac{b^4}{256a^3} = 0 \quad (a \neq 0) \right. \tag{10a}$$

(10)

$$x = -\frac{b}{4a} \mp \frac{\rho}{4a}\cos\left[\frac{1}{2}\cos^{-1}\left(\frac{16a^2C - \rho^3}{\rho^3}\right)\right] \pm \frac{1}{4a}\sqrt{\rho^2\left(1 \pm \cos\left[\frac{1}{2}\cos^{-1}\left(\frac{16a^2C - \rho^3}{\rho^3}\right)\right]\right)^2 - \frac{\rho^3 + 16a^2C + 8aB\rho}{\rho}} \tag{10b}$$

where ρ is found in (9c).

2. Replacing $B + \frac{3b^2}{8a}$ with c or $B = c - \frac{3b^2}{8a}$ in (10) gives

$$\left\{ ax^4 + bx^3 + cx^2 + \left(C - \frac{b^3}{8a^2} + \frac{bc}{2a}\right)x + D + \frac{Cb}{4a} - \frac{5b^4}{256a^3} + \frac{b^2c}{16a^2} = 0 \quad (a \neq 0) \right. \tag{11a}$$

$$x_{1,2,3,4} = -\frac{b}{4a} \mp \frac{\rho}{4a}\cos\left[\frac{1}{2}\cos^{-1}\left(\frac{16a^2C - \rho^3}{\rho^3}\right)\right] \tag{11}$$

$$\pm \frac{1}{4a}\sqrt{\rho^2\left(1 \pm \cos\left[\frac{1}{2}\cos^{-1}\left(\frac{16a^2C - \rho^3}{\rho^3}\right)\right]\right)^2 - \frac{\rho^3 + 16a^2C + \rho(8ac - 3b^2)}{\rho}} \tag{11b}$$

where

$$\rho = \frac{(8ac - 3b^2)^2 - 256a^3D}{96a^2C}$$

$$+\frac{\sqrt{\left(8ac-3b^2\right)^4+65536\,a^6D^2+3072\,a^4\left(8ac-3b^2\right)C^2-512\,a^3\left(8ac-3b^2\right)^2D}}{48\,a^2C}\cos\left(\frac{1}{3}\cos^{-1}\left(\frac{Q}{S}\right)\right)\quad(11c)$$

$$Q=\left(8ac-3b^2\right)^6+4608\,a^4\left(8ac-3b^2\right)^3C^2+3538944\,a^8C^4+196608\,a^6\left(8ac-3b^2\right)^2D^2$$
$$-768\,a^3\left(8ac-3b^2\right)^4D-1179648\,a^7\left(8ac-3b^2\right)C^2D-16777216\,a^9D^3$$

and

$$S=\left(\left(8ac-3b^2\right)^4+65536\,a^6D^2+3072\,a^4\left(8ac-3b^2\right)C^2-512\,a^3\left(8ac-3b^2\right)^2D\right)^{3/2}.$$

3. By replacing $C-\dfrac{b^3}{8a^2}+\dfrac{bc}{2a}$ with d or $C=d+\dfrac{b^3}{8a^2}-\dfrac{bc}{2a}$ in (11), which gives

$$\begin{cases} ax^4+bx^3+cx^2+dx+D+\dfrac{3b^4}{256a^3}-\dfrac{b^2c}{16a^2}+\dfrac{bd}{4a}=0 \quad (a\neq0) & (12a)\\[2em] x_{1,2,3,4}=\dfrac{-b\mp\rho}{4a}\cos\left[\dfrac{1}{2}\cos^{-1}\left(\dfrac{2\left(8a^2d+b^3-4abc\right)-\rho^3}{\rho^3}\right)\right] & (12)\\[2em] \pm\dfrac{1}{4a}\sqrt{\rho^2\left(1\pm\cos\left[\dfrac{1}{2}\cos^{-1}\left(\dfrac{2\left(8a^2d+b^3-4abc\right)-\rho^3}{\rho^3}\right)\right]\right)^2-\dfrac{\rho^3+2\left(8a^2d+b^3-4abc\right)+\rho\left(8ac-3b^2\right)}{\rho}} & (12b) \end{cases}$$

where

$$\rho=\frac{\left(8ac-3b^2\right)^2-256\,a^3D}{12\left(8a^2d+b^3-4abc\right)}$$

$$+\frac{\sqrt{\left(8ac-3b^2\right)^4+65536a^6D^2+48\left(8ac-3b^2\right)\left(8a^2d+b^3-4abc\right)^2-512a^3\left(8ac-3b^2\right)^2D}}{6\left(8a^2d+b^3-4abc\right)}\cos\left(\frac{1}{3}\cos^{-1}\left(\frac{Q}{S}\right)\right)\quad(12c)$$

$$Q=\left(8ac-3b^2\right)^6+72\left(8ac-3b^2\right)^3\left(8a^2d+b^3-4abc\right)^2+864\left(8a^2d+b^3-4abc\right)^4$$
$$+196608\,a^6\left(8ac-3b^2\right)^2D^2-768\,a^3\left(8ac-3b^2\right)^4D-18432\,a^3\left(8ac-3b^2\right)\left(8a^2d+b^3-4abc\right)^2D$$
$$-16777216\,a^9D^3-768\,a^3\left(8ac-3b^2\right)^4D-18432\,a^3\left(8ac-3b^2\right)\left(8a^2d+b^3-4abc\right)^2D-16777216\,a^9D^3$$

and

$$S=\left(\left(8ac-3b^2\right)^4+65536\,a^6D^2+48\left(8ac-3b^2\right)\left(8a^2d+b^3-4abc\right)^2-512\,a^3\left(8ac-3b^2\right)^2D\right)^{3/2}$$

4. Lastly, by replacing $D+\dfrac{3b^4}{256a^3}-\dfrac{b^2c}{16a^2}+\dfrac{bd}{4a}$ with e or $D=e-\dfrac{3b^4}{256a^3}+\dfrac{b^2c}{16a^2}-\dfrac{bd}{4a}$ in (12), which yields the solution to the general quartic equation:

$$ax^4+bx^3+cx^2+dx+e=0 \quad (a\neq0) \tag{13a}$$

$$x_{1,2,3,4}=-\frac{b}{4a}\mp\frac{\rho}{4a}\cos\left[\frac{1}{2}\cos^{-1}\left(\frac{2(8a^2d+b^3-4abc)-\rho^3}{\rho^3}\right)\right]$$

$$\pm\frac{1}{4a}\sqrt{\rho^2\left(1\pm\cos\left[\frac{1}{2}\cos^{-1}\left(\frac{2(8a^2d+b^3-4abc)-\rho^3}{\rho^3}\right)\right]\right)^2-\frac{\rho^3+2(8a^2d+b^3-4abc)+\rho(8ac-3b^2)}{\rho}} \tag{13b}$$

where $\tag{13}$

$$\left|\frac{2(8a^2d+b^3-4abc)-\rho^3}{\rho^3}\right|\leq1,$$

$$\rho=\frac{3b^4+16a^2bd+16a^2c^2-64a^3e-16ab^2c}{3(8a^2d+b^3-4abc)}+\frac{\sqrt{R}}{6(8a^2d+b^3-4abc)}\cos\left(\frac{1}{3}\cos^{-1}\left(\frac{Q}{S}\right)\right), \tag{13c}$$

$$\left|\frac{Q}{S}\right|\leq1,$$

$$R=(8ac-3b^2)^4+48(8ac-3b^2)(8a^2d+b^3-4abc)^2$$
$$+(256a^3e+112ab^2c-21b^4-64a^2bd-128a^2c^2)(256a^3e+16ab^2c-3b^4-64a^2bd),$$

$$Q=(8ac-3b^2)^6+72(8ac-3b^2)^3(8a^2d+b^3-4abc)^2+864(8a^2d+b^3-4abc)^4$$
$$+3(8ac-3b^2)^2(256a^3e+16ab^2c-3b^4-64a^2bd)^2$$
$$-3(8ac-3b^2)^4(256a^3e+16ab^2c-3b^4-64a^2bd)$$
$$-72(8ac-3b^2)(8a^2d+b^3-4abc)^2(256a^3e+16ab^2c-3b^4-64a^2bd)$$
$$-(256a^3e+16ab^2c-3b^4-64a^2bd)^3,$$

$$S=\left[(8ac-3b^2)^4+(256a^3e+16ab^2c-3b^4-64a^2bd)^2+48(8ac-3b^2)(8a^2d+b^3-4abc)^2\right.$$
$$\left.-2(8ac-3b^2)^2(256a^3e+16ab^2c-3b^4-64a^2bd)\right]^{3/2}$$
$$=4096a^3\left[256a^4e^2+96a^3cd^2-128a^3bde-128a^3c^2e+128a^2b^2ce+16a^2c^4\right.$$
$$\left.-20a^2b^2d^2-64a^2bc^2d+28ab^3cd-8ab^2c^3-24ab^4e+b^4c^2-3b^5d\right]^{3/2}.$$

Basic Analysis of Solution Formula (13)

+> If $\left|\dfrac{2(8a^2d+b^3-4abc)-\rho^3}{\rho^3}\right|\geq1,$

$$x_{1,2,3,4} = -\frac{b}{4a} \mp \frac{\rho}{4a} \cosh\left[\frac{1}{2}\cosh^{-1}\left(\frac{2(8a^2d+b^3-4abc)-\rho^3}{\rho^3}\right)\right]$$
$$\pm \frac{1}{4a}\sqrt{\rho^2\left(1\pm\cosh\left[\frac{1}{2}\cosh^{-1}\left(\frac{2(8a^2d+b^3-4abc)-\rho^3}{\rho^3}\right)\right]\right)^2 - \frac{\rho^3+2(8a^2d+b^3-4abc)+\rho(8ac-3b^2)}{\rho}}$$

$$(14a)$$

+> If $\left|\dfrac{2(8a^2d+b^3-4abc)-\rho^3}{\rho^3}\right| \le 1$, the roots to the quartic equation are

$$x_{1,2,3,4} = -\frac{b}{4a} \mp \frac{\rho}{4a} \cos\left[\frac{1}{2}\cos^{-1}\left(\frac{2(8a^2d+b^3-4abc)-\rho^3}{\rho^3}\right)\right]$$
$$\pm \frac{1}{4a}\sqrt{\rho^2\left(1\pm\cos\left[\frac{1}{2}\cos^{-1}\left(\frac{2(8a^2d+b^3-4abc)-\rho^3}{\rho^3}\right)\right]\right)^2 - \frac{\rho^3+2(8a^2d+b^3-4abc)+\rho(8ac-3b^2)}{\rho}}$$

$$(14b)$$

where

$$\begin{cases} \left|\dfrac{Q}{S}\right| > 1 \\[2mm] \rho = \dfrac{3b^4+16a^2bd+16a^2c^2-64a^3e-16ab^2c}{3(8a^2d+b^3-4abc)} + \dfrac{\sqrt{R}}{6(8a^2d+b^3-4abc)}\cosh\left(\dfrac{1}{3}\cosh^{-1}\left(\dfrac{Q}{S}\right)\right) \end{cases} \quad (14c)$$

or

$$\begin{cases} \left|\dfrac{Q}{S}\right| \le 1 \\[2mm] \rho = \dfrac{3b^4+16a^2bd+16a^2c^2-64a^3e-16ab^2c}{3(8a^2d+b^3-4abc)} + \dfrac{\sqrt{R}}{6(8a^2d+b^3-4abc)}\cos\left(\dfrac{1}{3}\cos^{-1}\left(\dfrac{Q}{S}\right)\right) \end{cases} \quad (14d)$$

(To explicitly derive solution if Q/S<-1, see more details in Section II for solving general cubic equation.)

$$R = (8ac-3b^2)^4 + 48(8ac-3b^2)(8a^2d+b^3-4abc)^2$$
$$+ (256a^3e+112ab^2c-21b^4-64a^2bd-128a^2c^2)(256a^3e+16ab^2c-3b^4-64a^2bd)$$

$$Q = (8ac-3b^2)^6 + 72(8ac-3b^2)^3(8a^2d+b^3-4abc)^2 + 864(8a^2d+b^3-4abc)^4$$
$$+ 3(8ac-3b^2)^2(256a^3e+16ab^2c-3b^4-64a^2bd)^2$$

$$-3\left(8ac-3b^2\right)^4\left(256a^3e+16ab^2c-3b^4-64a^2bd\right)$$

$$-72\left(8ac-3b^2\right)\left(8a^2d+b^3-4abc\right)^2\left(256a^3e+16ab^2c-3b^4-64a^2bd\right)$$

$$-\left(256a^3e+16ab^2c-3b^4-64a^2bd\right)^3$$

and

$$S=\left[\left(8ac-3b^2\right)^4+\left(256a^3e+16ab^2c-3b^4-64a^2bd\right)^2+48\left(8ac-3b^2\right)\left(8a^2d+b^3-4abc\right)^2\right.$$

$$\left.-2\left(8ac-3b^2\right)^2\left(256a^3e+16ab^2c-3b^4-64a^2bd\right)\right]^{3/2}$$

$$=4096a^3\left[256a^4e^2+96a^3cd^2-128a^3bde-128a^3c^2e+128a^2b^2ce+16a^2c^4\right.$$

$$\left.-20a^2b^2d^2-64a^2bc^2d+28ab^3cd-8ab^2c^3-24ab^4e+b^4c^2-3b^5d\right]^{3/2}.$$

[2-A-13]
Solution to Depressed Quartic Equation with Renamed Coefficients:

$$ax^4+bx^2+cx+d=0 \quad (a\neq0)$$

$$x_{1,2}=-\rho-\frac{4\sqrt{a}\left(\rho\right)^{3/2}\mp\sqrt{2}c}{4\sqrt{a\rho}}\left(1-\cos\left[\frac{1}{2}\cos^{-1}\left(1-\frac{8\left(c+4a\rho^3+2b\rho\right)}{\left(4\sqrt{a}\left(\rho\right)^{3/2}\mp\sqrt{2}c\right)^2}\right)\right]\right), \quad (95a)$$

$$x_{3,4}=-\rho-\frac{4\sqrt{a}\left(\rho\right)^{3/2}\mp\sqrt{2}c}{4\sqrt{a\rho}}\left(1+\cos\left[\frac{1}{2}\cos^{-1}\left(1-\frac{8\left(c+4a\rho^3+2b\rho\right)}{\left(4\sqrt{a}\left(\rho\right)^{3/2}\mp\sqrt{2}c\right)^2}\right)\right]\right), \quad (95b) \quad (95)$$

where

$$\rho=-\frac{4ad-b^2}{6ac}-\frac{\sqrt{\left(b^2-4ad\right)^2+6abc^2}}{3ac}*\cos\left(\frac{1}{3}\cos^{-1}\left(\frac{18abc^2\left(4ad-b^2\right)+2\left(4ad-c^2\right)^3-27a^2c^4}{2\left(\left(b^2-4ad\right)^2+6abc^2\right)^{3/2}}\right)\right). \quad (95c)$$

[2-A-14]
Formula (281c) has been derived with the coefficient for the x^6 term, as shown in the results below.

Special septic equation:

$$ax^7 + bx^6 + cx^5 + dx^4 + \left(-\frac{20b^2c}{147a^2} + \frac{25c^2}{63a} - \frac{4\mu\sqrt{3}\,\sqrt{3b^2-7ac}}{\sqrt{3}7a} + \frac{4b\mu}{7a}\right)x^3$$

$$+\left[\frac{\sqrt{3}\mu^2}{4\sqrt{3b^2-7ac}} + \frac{25bc^2}{147a^2} + \frac{48b^5}{2401a^1} + \frac{b^2\mu}{7a^2}\right.$$

$$\left. -\frac{23(3b^2-7ac)^{5/2}}{21609\sqrt{3}a^4} - \frac{4\sqrt{3}b\mu\sqrt{3b^2-7ac}}{49a^2} - \frac{40b^3c}{343a^3} - \frac{c\mu}{21a}\right]x^2$$

$$+\left[\frac{2\sqrt{3}\mu(3b^2-7ac)^{3/2} - 12\sqrt{3}b^2\mu(3b^2-7ac)^{1/2} + 10b^2c^2 + 18b^3\mu}{1029a^3} + \frac{5c^3 - 6bc\mu}{441a^2}\right.$$

$$+\frac{\sqrt{3}b\mu^2}{14a(3b^2-7ac)^{1/2}} - \frac{46\sqrt{3}b(3b^2-7ac)^{5/2} + 1242b^6}{453790a^5} - \frac{30b^4c}{2401a^4} - \frac{\mu^2}{14a}\right]x$$

$$-\frac{2(3b^2-7ac)^{7/2}}{3176523\sqrt{3}a^6} - \frac{23b^2(3b^2-7ac)^{5/2}}{1058841\sqrt{3}a^6} + \frac{2b\mu(3b^2-7ac)^{3/2}}{2401\sqrt{3}a^4} - \frac{4b^3\mu(3b^2-7ac)^{1/2}}{2401\sqrt{3}a^4}$$

$$+\frac{\mu^2(3b^2-7ac)^{1/2}}{196\sqrt{3}a^2} - \frac{5\sqrt{3}\mu^2(3b^2-7ac)^{1/2}}{1372a^2} + \frac{\sqrt{3}b^2\mu^2}{196a^2(3b^2-7ac)^{1/2}} + \frac{50b^7}{823543a^6} - \frac{2b^5c}{50421a^5}$$

$$+\frac{423b^4\mu - 280b^3c^2}{302526a^4} + \frac{35bc^3 - 78b^2c\mu}{21609a^3} + \frac{19c^2\mu - 63b\mu^2}{6174a^2} + \frac{3\mu^3}{56(3b^2-7ac)} = 0$$

where $\mu = \dfrac{10b^3 + 49a^2d + 35abc}{49a^2}$.

Solution formula:

$$x_{k+1} = -\frac{b}{7a} + \frac{\sqrt{3b^2-7ac}}{\sqrt{3}7a} - \frac{\sqrt{3\sqrt{3}(10b^3+49a^2d+35abc) - 10(3b^2-7ac)^{3/2}}}{2\sqrt{42a}\sqrt[4]{3b^2-7ac}\sinh\left(\frac{1}{7}\sinh^{-1}\left(\frac{\sqrt{3\sqrt{3}(10b^3+49a^2d+35abc)-10(3b^2-7ac)^{3/2}}}{2\sqrt{2}[7a(3b^2-7ac)]^{3/4}}\right) + \frac{i2\pi k}{7}\right)}.$$

Or it can be expressed in terms of the sinht function to provide a complete solution,

$$x_{k+1} = -\frac{b}{7a} + \frac{\sqrt{3b^2-7ac}}{\sqrt{3}7a} - \frac{\sqrt{3\sqrt{3}(10b^3+49a^2d+35abc) - 10(3b^2-7ac)^{3/2}}}{2\sqrt{42a}\sqrt[4]{3b^2-7ac}\sinh_{7,k}\left(\frac{\sqrt{3\sqrt{3}(10b^3+49a^2d+35abc)-10(3b^2-7ac)^{3/2}}}{2\sqrt{2}[7a(3b^2-7ac)]^{3/4}}\right)},$$

for k = 0, 1, 2,...,6.

Author's Comments

➢ While we designate the chapter I with the title 'Nested Radicals and Algebraic Identities' and emphasize the exploration of nested radical and identities, it is crucial to underscore the significance of certain formulas, namely (1.10b), (1.2.9), (1.4.11), (1.5.9), and so forth. The integration of these formulas with others derived from definitions, such as (1.3.3b), (1.6.3), (1.13.17) and others, results in the development of groundbreaking advanced algebraic formulas, such as (1.3.11), (1.3.21), (1.3.28), (1.6.18), and so forth that transcend detailed verbal explication.

➢ The technique of solving polynomial equations using circular trigonometric functions or hyperbolic trigonometric functions possesses the feature of ensuring consistency in determining the entire solution once one of its solutions is found. This property is substantiated by the periodicity of the circular trigonometric function or hyperbolic trigonometric function being used. The number of roots of an equation is equivalent to the number of periodic indices associated with the trigonometric function utilized. The presence of trigonometric or hyperbolic trigonometric functions in the solution form is crucial in determining whether an equation possesses real solutions exclusively or a combination of both real and complex roots. To illustrate, when the solution of an equation is expressed as circular trigonometric functions, it signifies that the equation has purely real roots. In contrast, if the solution involves other types of hyperbolic trigonometric functions, it suggests the existence of a combination of real and complex roots. As we see the process of solving polynomial equations through trigonometric or hyperbolic trigonometric functions includes establishing a system of simultaneous two equations that are equal in value and solving them concurrently. One equation is shaped to construct a specific polynomial equation devoid of trigonometric or hyperbolic trigonometric functions by adjusting or replacing coefficients through addition, subtraction, multiplication, or division. The other equation represents the solution to the first equation and holds coefficients aligned with the equivalent math operations performed in the first equation. Ultimately, utilizing the properties encompassing periodicity, domain, and range inherent in each trigonometric or hyperbolic trigonometric function to determine the number of real and/or complex roots of the equation.

➢ Although this method can be time-consuming, the solutions to the polynomial equations are usually consistent. Even though the solution may be lengthy and complex, it is worth persevering until the end if it guarantees an answer. Due to the existence of numerous elementary trigonometric identities, we posit that they can be used to derive solution formulas for respective polynomial equations. We invite readers to apply the single substitution method outlined in this chapter for this purpose. Furthermore, various other elementary methods, such as employing multiple substitutions involving more than two unknown parameters, incorporating the Tschirnhaus transformation method for transforming a polynomial equation (which has not been utilized yet), or combining multiple solution formulas, remain unexplored.

➢ If we obtain solution to a polynomial equation in the form of $\cos\left((1/n)\cos^{-1}(x)\right)$ or $\cosh\left((1/n)\cosh^{-1}(x)\right)$, then the combination of these two forms are sufficient for determining roots of the equation across all x values. In fact, both forms represents a real algebraic expression for $-1 \leq x \leq 1$ and $1 \leq x < \infty$, but for x in the interval $(-\infty, -1]$, it is a complex expression.

➢ The newly defined functions, sinht, cosht and tanht, facilitate the computation of real roots of equations once they are implemented in calculators. These functions not only represent the roots of a polynomial equation but also serve as transitional functions, encompassing trigonometric and hyperbolic trigonometric functions. They establish a connection between algebra and trigonometry through polynomial equations. Upon integration as fundamental functions, they may help us understand the order of complex numbers seem to be deliberately arranged for each functional structure in mathematics that we must follow. The functional structure discussed here refers to the combination of both trigonometric and hyperbolic trigonometric functions.

➢ The formulas presented in section 'Transforming Solution Formulas from Trigonometric or Hyperbolic Trigonometric Functions into Radical Forms' are crucial because they can be utilized to transform all solution formulas derived in this chapter into radical forms.

➢ Generally, this approach necessitates not only a fundamental comprehension of algebra but also knowledge of trigonometry and hyperbolic trigonometry. Delving into these subjects enhances our comprehension of the realm of complex numbers by establishing an interactive link between trigonometric functions and hyperbolic trigonometric functions. Ultimately, it provides a comprehensive overview of the intricate interrelationship among Algebra, Trigonometry, and complex numbers, all encapsulated within polynomial equations and their solution formulas.

➢ One of the particularly noteworthy results is equation (356a), as it sheds light on the feasibility of solve other higher degree equations in terms of radicals. This is achievable when e/c = 'constant' such that either cx^5 or ex^3 term must be disappeared or eliminated from the equation.

➢ Some tips:
  → Considering the following a simple situation: Assuming m be any number and i is imaginary unit, then $i\sqrt{-m}=i \cdot i\sqrt{m}=i^2\sqrt{m}=-\sqrt{m}$ is correctly following the algebraic rule. However, if we deviate from this rule, it can be demonstrated expanding i that $i\sqrt{-m}=\sqrt{-1}\sqrt{-m}=\sqrt{(-1)(-m)}=\sqrt{m}$. The result appears to be contradictory or confusing, but what can we learn from this situation? When the solution formula contains radicals in the principal branch, it should not

be simplified immediately in some rare scenarios that lead to contradict result. Instead, specific conditions for the variable m, for example or its connection with other expressions are necessary to ensure that both satisfy a derived value. This is why for each step-by-step we do not consider sign of any coefficients that may be within radical(s), and instead carefully examine them after we completely derive the solution formula. This approach helps us avoid getting trapped in an endless cycle where results are uncertain.

→ Individuals who engage in the pursuit of solution formulas for polynomial equations using trigonometric and hyperbolic trigonometric functions over time will experience significant improvement in their mathematical skills, particularly in Algebra, Trigonometry, and Complex Number.

➢ **Author's Mischievous Note**

We leave a Machin-like formula without divulging its specifics in this book. The formula likes a mischievous note as our clandestine timestamp and is expressed as

$$\tan^{-1}\left(\frac{1}{2}\right)+\tan^{-1}\left(\frac{1}{20}\right)+\tan^{-1}\left(\frac{1}{2024}\right)-\frac{1}{4}\tan^{-1}\left(\frac{31488243646751629927}{59664553872995095464}\right)=\frac{\pi}{8}.$$

As winter bids adieu on February 20, 2024, remind that: Mathematics, much like love, flourishes in secrecy. It sparks curiosity, which in turn may lead to intricate journeys of exploration.

— Tue Vu

References

[1] Earl W. Swokowski (1984). Calculus with Analytic Geometry. Third Edition. PWS Publishers. P. 388, 391, 394.

[2] Inverse hyperbolic functions, https://en.wikipedia.org/wiki/Inverse_hyperbolic_functions. From Wiki.

[3] Berndt, Bruce C. Ramanujan's Notebooks Part IV. Springer-Verlag New York, Inc., 1994, p. 31.

[4] Raymond A. Barnett (1984). College Algebra, Trigonometry, and Analytic Geometry. Third Edition. McGraw-Hill. P 490.

[5] Raymond A. Barnett (1984). College Algebra, Trigonometry, and Analytic Geometry. Third Edition. McGraw-Hill. P 495, 497.

[6] Lemniscate elliptic functions, https://en.wikipedia.org/wiki/Lemniscate_elliptic_functions, from Wiki.

[7] Raymond A. Barnett (1984). College Algebra, Trigonometry, and Analytic Geometry. Third Edition. McGraw-Hill. P 123.

[7] Quadratic equation, https://en.wikipedia.org/wiki/Quadratic_equation. From Wiki.

[8] Spiegel, M. R., Lipschutz, S., & Liu, J. (2009). Mathematical Handbook of Formulas and Tables. Third Edition. McGraw-Hill. P 13.

[8] Cubic Equation, https://en.wikipedia.org/wiki/Cubic_equation. From Wikipedia.

[9] Spiegel, M. R., Lipschutz, S., & Liu, J. (2009). Mathematical Handbook of Formulas and Tables. Third Edition. McGraw-Hill. P 14.

[9] Quartic Equation, https://en.wikipedia.org/wiki/Quartic_equation. From Wikipedia, the free encyclopedia.

[10] J. J. Sylvester (1851) "On a remarkable discovery in the theory of canonical forms and of hyper determinants," Philosophical Magazine, 4th series, 2: 391–410; Sylvester coined the term "discriminant" on page 406.

[11] Weisstein, Eric W. "de Moivre's Quintic." From MathWorld--A Wolfram Web Resource. https://mathworld.wolfram.com/deMoivresQuintic.html

[12] OEIS Foundation Inc. (2024), Triangle of coefficients of Chebyshev polynomials, Entry A084930 in The On-Line Encyclopedia of Integer Sequences, https://oeis.org/A084930.

www.ingramcontent.com/pod-product-compliance
Lightning Source LLC
Chambersburg PA
CBHW061617210326
41520CB00041B/7476